NON-CHEMICAL WEED CONTROL

NON-CHEMICAL WEED CONTROL

Edited by

KHAWAR JABRAN
*Department of Plant Protection, Faculty of Agriculture and Natural Sciences,
Duzce University, Duzce, Turkey*

BHAGIRATH S. CHAUHAN
*The Centre for Plant Science, Queensland Alliance for Agriculture and Food Innovation (QAAFI),
The University of Queensland, Australia*

ACADEMIC PRESS
An imprint of Elsevier

Academic Press is an imprint of Elsevier
125 London Wall, London EC2Y 5AS, United Kingdom
525 B Street, Suite 1800, San Diego, CA 92101-4495, United States
50 Hampshire Street, 5th Floor, Cambridge, MA 02139, United States
The Boulevard, Langford Lane, Kidlington, Oxford OX5 1GB, United Kingdom

Notices
Knowledge and best practice in this field are constantly changing. As new research and experience broaden our understanding, changes in research methods, professional practices, or medical treatment may become necessary.

Practitioners and researchers must always rely on their own experience and knowledge in evaluating and using any information, methods, compounds, or experiments described herein. In using such information or methods they should be mindful of their own safety and the safety of others, including parties for whom they have a professional responsibility.

To the fullest extent of the law, neither the Publisher nor the authors, contributors, or editors, assume any liability for any injury and/or damage to persons or property as a matter of products liability, negligence or otherwise, or from any use or operation of any methods, products, instructions, or ideas contained in the material herein.

Library of Congress Cataloging-in-Publication Data
A catalog record for this book is available from the Library of Congress

British Library Cataloguing-in-Publication Data
A catalogue record for this book is available from the British Library

ISBN 978-0-12-809881-3

For information on all Academic Press publications
visit our website at https://www.elsevier.com/books-and-journals

Working together
to grow libraries in
developing countries

www.elsevier.com • www.bookaid.org

Publisher: Andre Wolff
Acquisition Editor: Nancy Maragioglio
Editorial Project Manager: Billie Jean Fernandez
Production Project Manager: Priya Kumaragururparan
Cover Designer: Billie Jean Fernandez

Typeset by SPi Global, India

Contents

Contributors

Mehmet Arslan Erciyes University, Kayseri, Turkey

Graham Brodie Melbourne University, Dookie, VIC, Australia

Bhagirath S. Chauhan The University of Queensland, Gatton, QLD, Australia

Shahid Farooq Faculty of Agriculture, Gaziosmanpaşa University, Tokat, Turkey

Mubshar Hussain Bahauddin Zakariya University Multan, Multan, Pakistan

Khawar Jabran Düzce University, Düzce, Turkey

Nicholas E. Korres University of Arkansas, Fayetteville, AR, United States

Carolyn J. Lowry University of New Hampshire, Durham, NH, United States

Charles Merfield The BHU Future Farming Centre, Canterbury, New Zealand

Arslan M. Peerzada The University of Queensland, Gatton, QLD, Australia

Richard G. Smith University of New Hampshire, Durham, NH, United States

Ahmet Uludag Çanakkale Onsekiz Mart University, Çanakkale, Turkey; Düzce University, Düzce, Turkey

Ilhan Uremis Mustafa Kemal University, Hatay, Turkey

Preface

Weeds are the yield-limiting biotic factors that grow in crops without human attention. They feed on moisture and nutrients, which are otherwise to be used by crop plants. Also, weeds occupy space and snatch the light that is supposed to be utilized by crop plants. In addition, weeds interfere with the crop growth processes by exuding certain chemicals and also through root interferences. Ultimately, weeds disturb the growth of crop plants, deprive them of resources, decrease the crop yields, and increase the cost of production. In some crops, weeds may cause minor yield losses $\sim$<10%; however, in several field crops (having an important role in global food security), weeds may reduce yield by nearly 50%. In several instances, weeds can also result in total failure of the crop. On a global scale, these losses approach to billions of dollars along with a considerable reduction in food availability. Owing to these facts, weed researchers are working all over the world to fight against weeds.

Since its advent in the mid-20th century, chemical control has been the most followed weed control method. This weed control method contributed significantly to reduce the damages caused by weeds to crops and highly improved the crop yields over the last 5–6 decades. However, several problems are also associated with this weed control method. Most important of these include herbicide residues in food, environmental pollution, and herbicide resistance evolution in weeds. Hence, researchers from all across the globe are focusing on weed control

methods that may provide an alternative to herbicidal weed control. Moreover, chemical weed control is not permitted in cropping systems such as organic farming. Hence, nonchemical weed control methods are inevitable to suppress weeds in such farming systems.

Classical nonchemical weed control methods include the use of preventive, cultural, and mechanical weed control methods. Although these methods possess effectiveness against weeds, they do not provide enough weed control, which may satisfy the standards of intensive and precision farming. Recent literature addresses the use of nonchemical weed control methods in modern agriculture and organic farming. However, in most of the cases, the classical techniques of nonchemical weed control have been discussed. Moreover, most of such literature is composed of research publications rather than books or monographs. This book brings the most recent nonchemical weed control techniques (in addition to classical nonchemical weed control methods). The book is useful for teaching and research staff of universities and agricultural research institutes and students. Other communities who will benefit from this book include the extension specialists and growers, particularly the ones who prefer to grow their food organically.

Khawar Jabran
Duzce University, Turkey

Bhagirath S. Chauhan
The University of Queensland, Australia

1

Overview and Significance of Non-Chemical Weed Control

Khawar Jabran, *Bhagirath S. Chauhan*[†]

*Düzce University, Düzce, Turkey [†]The University of Queensland, Gatton, QLD, Australia

1.1 INTRODUCTION

In addition to other criteria, effective control of all pests is required to achieve sustainability in any agricultural system. Weeds are a serious pest of crops and have been infesting crops since the onset of agriculture. Optimum growth and development of crop plants depend on the balanced nutrition and availability of moisture, space, and photosynthetically active radiation (Fageria et al., 2010). Weeds also aim at consuming these resources for their growth, development, and reproduction. Hence, weeds and crop plants are usually in a consistent competition for nutrition, space, moisture, and light (Guglielmini et al., 2016). Weeds also interfere with the growth of crop plants, for instance, they may reduce the light reaching crop plants through shading or exude some allelochemicals that impact crop plants negatively. Hence, a sustainable weed control is required owing to severe weed-crop competition and damages caused by weeds to crops.

Throughout the human history, physical or mechanical methods of weed control have been used. It was the end of the nineteenth century when some inorganic salts (e.g., copper salts) and botanical oils were used as pesticides for controlling weeds or other pests. Discovery and introduction of 2,4-D was an important step that opened the doors to the discovery, development, and formulation of several other herbicides (Peterson, 1967). After their discovery and use in agriculture, herbicides have been providing a great contribution in managing weeds in increasing crop yields.

Despite all the benefits gained through herbicide application, resistance evolution in weeds against herbicides and environmental concerns have come out as serious issues (Powles, 2008; Heap, 2014). The sustainability of weed control achieved through herbicide application has become questionable after the onset of issues of herbicide resistance,

1

environmental pollution, and other ecological concerns (Powles, 2008; Heap, 2014; Gaba et al., 2016). In the wake of these issues, non-chemical weed control methods are required to substitute or at least supplement the herbicidal weed control. The European Union (EU) advises the growers to adopt integrated pest management while banning many of herbicides due to environmental implications of their use (Moss, 2010).

Non-chemical weed control may provide the benefits such as safe and healthy food production and environmental conservation. Such weed control methods are also inevitable for the successful implementation of organic farming systems (Bond and Grundy, 2001; Jabran et al., 2015). The current situation, however, indicates that knowledge gaps exist regarding the technologies involved in non-chemical weed control methods. This book *Non-Chemical Weed Control* is aimed at bringing together various important non-chemical weed control methods. The book is also aimed at discussing the knowledge gaps and flaws in the use of non-chemical methods for weed management and the possible solutions.

This chapter is aimed at providing a brief introduction to various concepts and issues that are related to non-chemical weed control. The chapter discusses the factors that highlight the need for non-chemical weed control, the relationship of food security and weed control, and finally, provides an overview of non-chemical weed control techniques.

1.2 HERBICIDE-RESISTANT WEEDS

Resistant evolution in weeds against herbicides is the greatest challenge faced by weed science (Heap, 2014). There are >250 weed species that have evolved resistance to 161 herbicides; in total, there are 26 known herbicide sites of actions, and weeds have evolved resistance against 23 of these (http://www.weedscience.org/). Out of the resistant weeds, >100 are narrow-leaved, while nearly 150 are broad-leaved plant species. The number of countries and crops where herbicide-resistant weeds have been reported is 68 and 91, respectively.

The problem of herbicide resistance in weeds has been worsened due to factors such as a limited number of existing mode of actions and a rare discovery of new herbicides or target sites. Moreover, repeated use of same herbicides and monoculture accelerates the herbicide resistance evolution in weeds. Non-chemical becomes the alternate weed control strategy in the wake of the existence of herbicide resistance in weeds.

1.3 ENVIRONMENTAL AND HEALTH CHALLENGES

One among the aims of sustainable development is the provision of a safe environment to current and future generations. Environmental safety may be compromised and harmed by the overuse or misuse of chemicals, such as herbicides. The houses located in a close proximity (<750 m) to farming fields with the heavy use of herbicides are easily contaminated with herbicide residues (Ward et al., 2006). Herbicide application is aimed at controlling weeds in crop plants; however, nontarget plant species are also frequently damaged after spray. Recent work from France indicated that applications of herbicides were killing the rare plant species while having no positive effect on crop yields or control of problematic weeds in wheat (Gaba et al., 2016). These results show there can be certain ecological issues as a result of herbicide use.

The negative impacts of herbicides on environment or organisms have been reported with scientific evidence. For example, atrazine was found to disturb reproduction and hermaphroditism in wild leopard frogs (*Rana pipiens*) (Hayes et al., 2002). Rohr and Palmer (2005) evaluated the sensitivity of salamanders *Ambystoma barbouri* to various concentrations of atrazine herbicides. An exposure of salamanders to ≥ 40 µg/L of this herbicide changed their behavior and accelerated a water loss in their body.

Not only honeybees facilitate the pollination in many plants, but also their visits may improve the yield and quality of many fruit and field crops. Recently, a decrease in honeybee colonies has been observed in areas (particularly, in the United States), where pesticides were used at a high rate (Potts et al., 2016). Herbicides are also expected to cause serious adverse effects on honeybees. For example, exposure of honeybees to atrazine, glyphosate, or their combinations could disturb their metabolic pathway (Jumarie et al., 2017).

Sterling and Arundel (1986) reviewed the effects of phenoxy herbicides on health and reported that these herbicides might cause several diseases and illnesses in humans. They noted that phenoxy herbicides were associated with congenital abnormalities (if parents were exposed to these herbicides) and various types of cancers. Similarly, heavy use of glyphosate has increased the chances of human exposure to this herbicide during and after application under field conditions and through consumption of water contaminated with glyphosate (Myers et al., 2016). A study conducted in Sri Lanka indicated that kidney diseases in farmers were associated with spraying glyphosate and other pesticides in paddy fields without wearing suitable clothes and other protections and drinking water from an abandoned well that had been previously sprayed with glyphosate (Jayasumana et al., 2015). According to Camacho and Mejia (2015), aerial sprays of glyphosate in Colombia caused several health problems to humans living in that area. Most important among health issues were an increased number of miscarriages, increased number of respiratory and skin diseases, and increased visit number of patients for availing medical care. Research from the US studied relations of acetochlor to cancer (Lebov et al., 2015). There was a possible link between exposure to acetochlor and cancer; however, the researchers did not provide a firm conclusion owing to lack of clear trends in results. In a study from France, high concentrations of acetochlor and alachlor were recorded in the urine of pregnant women holding rural background (Chevrier et al., 2014).

1.4 FOOD SECURITY AND WEED CONTROL

Global food security is directly dependent on the amount of cereals, pulses, vegetables, and fruits that are produced from the agricultural fields. A decrease in the production of these food types will disturb the supply of food items (Alexandratos and Bruinsma, 2012). Hence, the factors that constrain the realization of production potential of food crops indirectly disturb the food security (Kahane et al., 2013). This implies that food security is indirectly negatively impacted by several abiotic and biotic factors that hinder the crop productivity.

Weed infestation can impact the food security by decreasing the crop productivity in infested fields. For instance, a heavy weed infestation may impact food securities in the countries where a major portion of food needs are fulfilled through single or a few crops (Tshewang et al., 2016). Weeds cause a decrease in crop yield and produce quality as well. Compared with other pests, weeds are known to cause the highest damage to crop productivity (Oerke, 2006). For instance, weeds may cause 35% decrease in productivity of cotton, 37% in rice and soybeans, 30% in potatoes, 40% in maize, and 23% in wheat (Oerke, 2006). The lack of weed control or inefficient weed management will cause a severe decline in productivity of crops that will ultimately disturb the regional and global food security. Hence, an effective weed control is required to ensure adequate food production and food security.

1.5 ORGANIC CROP PRODUCTION AND WEED CONTROL

Organic production systems are advocated owing to their health benefits achieved through the provision of food that is free of harmful chemical compounds. Organic crop production systems do not allow the use of synthetic herbicides for weed control. Hence, non-chemical weed control is required to be practiced under such conditions (Bond and Grundy, 2001). Prohibition of herbicide use in organic agriculture results in high weed infestations compared with the fields under conventional agriculture. Under organic crop production, weed management is totally relied on non-chemical methods, such as tillage, cover crops, high seed rates, and mulches. Demand for organic food will increase the importance of non-chemical weed control.

1.6 ROLE OF NON-CHEMICAL WEED CONTROL IN INTEGRATED WEED MANAGEMENT

Integrated weed management (IWM) makes use of multiple techniques to achieve sustainable weed control (Harker and O'Donovan, 2013). Use of IWM for controlling weeds can provide environmental and economic benefits. Use of herbicides is one among these techniques in IWM, while the rest comprise non-chemical practices. This means that achieving IWM is impossible without practicing the non-chemical weed control methods. There is a range of choices for non-chemical weed control methods that can be chosen according to the nature of crop, intensity of weed infestation, climate, growth stages of weeds and crops, critical period for crop-weed competition, available resources, and yield goals (Harker and O'Donovan, 2013; Bajwa et al., 2015). Advancement in non-chemical weed control is inevitable if sustainable IWM needs to be achieved.

1.7 ROLE OF NON-CHEMICAL WEED CONTROL IN HOME LAWNS, HOME GROWN VEGETABLES

The location of home lawns and home gardens is usually very close to residential places; hence, it may not be a good idea to apply herbicides in these areas for controlling weeds.

There is greater likeliness of human contact to herbicides from these areas compared with agricultural fields outside (Harris and Solomon, 1992; Nishioka et al., 1996).

The vegetables are consumed fresh or cooked almost immediately after their harvest, whereas cereal grains, legumes, etc. are consumed much later after herbicide application. This long duration may cause a decay/loss of herbicides in cereals or legumes; however, this may not be possible in short periods available with vegetables (Tadeo et al., 2000; Claeys et al., 2011). Many times, herbicide spray may directly fall on fruits if the weeds are sprayed on fruit-bearing vegetable plants. These factors make non-chemical weed control more important for home lawns or home-grown vegetables. Ultimately, replacing herbicide application with non-chemical weed control under such places will help to achieve environmental protection and health safety.

1.8 AN OVERVIEW OF RECENT NON-CHEMICAL WEED CONTROL METHODOLOGIES

All of the ancient weed control techniques were non-chemical. It was until the late 19th century when some inorganic chemicals were used for controlling weeds. The discovery of 2,4-D in the middle of the 20th century started a new era of excellence in the discipline of weed control. Afterward, herbicides have been contributing tremendously in controlling weeds and increasing crop productivity. Even today, herbicide application is important, but not a sole method of weed control. Non-chemical weed control methods may be complex, possess a lower efficacy than herbicides, and require high expenditure, but they are needed for the sake of environmental and health safety (Moss, 2010).

Weed control through herbicide application is usually supplemented with non-chemical weed control methods. Broad categories of non-chemical weed control may include preventive strategies, cultural weed control, physical/mechanical weed control, allelopathy, and biological weed control. Table 1.1 provides a summary of currently available non-chemical weed control methods. Weed control methods such as the use of tillage have usually been a consistent constituent of weed management plans for farmers. Although less frequent, there has been an implementation of preventive measures and classical cultural weed control (such as crop rotation, intercropping, and cover cropping). The cultural technique of flooding may be useful for controlling weeds in rice. Farmers always have a choice to observe such easy, inexpensive, and environment-friendly methods to suppress weeds at their farms. Usually, these methods may require some agronomic management and negligible extra costs but could help effectively in reducing weed populations in various crop production systems.

In non-chemical weed control, most importantly, farmers have the choice to choose the competitive crop cultivars against weeds and increase the pressure on weeds through agronomic management (e.g., high seed rates and narrow row spacing). The science of allelopathy provides an important option to suppress weeds by sowing allelopathic crop cultivars (Jabran et al., 2015; Jabran, 2017). Like allelopathy, other nonconventional, non-chemical weed control methods are developing quickly. Important examples of such methods may include thermal weed control and use of electric systems and electromagnetic fields for suppressing weeds.

TABLE 1.1   A Brief Summary of Important Non-Chemical Weed Control Options Described in Previous Studies

Non-Chemical Weed Control Options	References
Prevention	Jabran et al. (2017)
Mulching	Crutchfield et al. (1986) and Jabran and Chauhan (2015)
Soil solarization, stale seedbed	Jabran and Chauhan (2015)
Flame weeding	Ascard (1995)
Hot water/steaming treatments	Hansson and Mattsson (2002)
Allelopathic/competitive cultivars	Sardana et al. (2017), Jabran et al. (2015), and Jabran (2017)
Row spacing, seed rate, row direction	Sardana et al. (2017)
Electric systems, electromagnetic fields	Diprose et al. (1984)
Allelopathy	Farooq et al. (2011), Jabran et al. (2015), and Jabran (2017)
Cover crops, intercopping	Jabran and Chauhan (2015)
Crop rotation	Shahzad et al. (2016)
Mechanical weed control (tillage)	Shaner and Beckie (2014)
Robotic weed control	Pérez-Ruíz et al. (2014)
Biological weed control	Winston et al. (2014)

Ground-cover systems (mulches, e.g., black polyethylene and straw mulch) are getting popularity as weed control methods, particularly under greenhouse conditions. In addition to regulating soil temperature, and, soil and water conservations, these techniques also suppress weeds. Over the time, there has been very slow progress in the development of biological control as an acceptable, effective, and sustainable weed management method. Processes involved in the synthesis of biological weed control agents and implementation of biological weed control have been a complex phenomenon. More investment, research, and practical demonstrations are required to make biological control an acceptable method of weed management. Robots and intelligent (tillage) implements are among the modern methods of weed control (Shaner and Beckie, 2014; Rueda-Ayala et al., 2015). Intelligent weeders can help in selective weed control by tilling only the parts of fields that are infested with weeds; only infested patches are tilled, while weed-free parts of fields are not tilled at all (Rueda-Ayala et al., 2015).

1.9 CONCLUSIONS

Relying on non-chemical weed control may help to achieve the goal of environment and health safety. Other than field crops, non-chemical weed control techniques may be utilized to suppress weeds in lawns and vegetable gardens, hence minimizing the chances of human exposure to herbicide remains. Further, non-chemical weed control will be useful in managing weeds that have evolved resistance to herbicides. Nevertheless, a variety of non-chemical weed control methods are available that could be selected according to specific field conditions for use in IWM.

References

Alexandratos, N., Bruinsma, J., 2012. World Agriculture Towards 2030/2050: The 2012 Revision, vol. 12-03. Rome, FAO, p. 4 (ESA Working Paper).

Ascard, J., 1995. Effects of flame weeding on weed species at different developmental stages. Weed Res. 35 (5), 397–411.

Bajwa, A.A., Mahajan, G., Chauhan, B.S., 2015. Nonconventional weed management strategies for modern agriculture. Weed Sci. 63, 723–747.

Bond, W., Grundy, A.C., 2001. Non-chemical weed management in organic farming systems. Weed Res. 41 (5), 383–405.

Camacho, A., Mejia, D., 2015. In: The health consequences of aerial spraying of illicit crops: the case of Colombia.Center for Global Development Working Paper No. 408. Available from SSRN: https://ssrn.Com/abstract=2623145 or https://doi.org/10.2139/ssrn.2623145 (accessed 09.03.17).

Chevrier, C., Serrano, T., Lecerf, R., Limon, G., Petit, C., Monfort, C., Hubert-Moy, L., Durand, G., Cordier, S., 2014. Environmental determinants of the urinary concentrations of herbicides during pregnancy: the PELAGIE mother-child cohort (France). Environ. Int. 63, 11–18.

Claeys, W.L., Schmit, J.F., Bragard, C., Maghuin-Rogister, G., Pussemier, L., Schiffers, B., 2011. Exposure of several Belgian consumer groups to pesticide residues through fresh fruit and vegetable consumption. Food Control 22, 508–516.

Crutchfield, D.A., Wicks, G.A., Burnside, O.C., 1986. Effect of winter wheat (*Triticum aestivum*) straw mulch level on weed control. Weed Sci. 34, 110–114.

Diprose, M.F., Benson, F.A., Willis, A.J., 1984. The effect of externally applied electrostatic fields, microwave radiation and electric currents on plants and other organisms, with special reference to weed control. Bot. Rev. 50 (2), 171–223.

Fageria, N.K., Baligar, V.C., Jones, C.A., 2010. Growth and Mineral Nutrition of Field Crops. CRC Press, New York.

Farooq, M., Jabran, K., Cheema, Z.A., Wahid, A., Siddique, K.H., 2011. The role of allelopathy in agricultural pest management. Pest Manag. Sci. 67 (5), 493–506.

Gaba, S., Gabriel, E., Chadœuf, J., Bonneu, F., Bretagnolle, V., 2016. Herbicides do not ensure for higher wheat yield, but eliminate rare plant species. Sci. Rep. 6, 30112. https://doi.org/10.1038/srep30112.

Guglielmini, A.C., Verdú, A.M.C., Satorre, E.H., 2016. Competitive ability of five common weed species in competition with soybean. Int. J. Pest Manag. 63, 30–36.

Hansson, D., Mattsson, J.E., 2002. Effect of drop size, water flow, wetting agent and water temperature on hot-water weed control. Crop. Prot. 21 (9), 773–781.

Harker, K.N., O'Donovan, J.T., 2013. Recent weed control, weed management, and integrated weed management. Weed Technol. 27, 1–11.

Harris, S.A., Solomon, K.R., 1992. Human exposure to 2,4-d following controlled activities on recently sprayed turf. J. Environ. Sci. Health B 27, 9–22.

Hayes, T., Haston, K., Tsui, M., Hoang, A., Haeffele, C., Vonk, A., 2002. Herbicides: feminization of male frogs in the wild. Nature 419, 895–896.

Heap, I., 2014. Global perspective of herbicide-resistant weeds. Pest Manag. Sci. 70 (9), 1306–1315.

Jabran, K., 2017. Manipulation of Allelopathic Crops for Weed Control. Springer International AG, Switzerland.

Jabran, K., Chauhan, B.S., 2015. Weed management in aerobic rice systems. Crop. Prot. 78, 151–163.

Jabran, K., Hussain, M., Chauhan, B.S., 2017. Integrated weed management in maize cultivation: an overview. In: Watson, D. (Ed.), Achieving Sustainable Cultivation of Maize. Burleigh Dodds Science Publishing Ltd., Cambridge, UK (In press).

Jabran, K., Mahajan, G., Sardana, V., Chauhan, B.S., 2015. Allelopathy for weed control in agricultural systems. Crop. Prot. 72, 57–65.

Jayasumana, C., Paranagama, P., Agampodi, S., Wijewardane, C., Gunatilake, S., Siribaddana, S., 2015. Drinking well water and occupational exposure to Herbicides is associated with chronic kidney disease, in Padavi-Sripura, Sri Lanka. Environ. Health 14, 6. https://doi.org/10.1186/1476-069X-14-6.

Jumarie, C., Aras, P., Boily, M., 2017. Mixtures of herbicides and metals affect the redox system of honey bees. Chemosphere 168, 163–170.

Kahane, R., Hodgkin, T., Jaenicke, H., Hoogendoorn, C., Hermann, M., Hughes, J.D.A., Padulosi, S., Looney, N., 2013. Agrobiodiversity for food security, health and income. Agron. Sustain. Dev. 33 (4), 671–693.

Lebov, J.F., Engel, L.S., Richardson, D., Hogan, S.L., Sandler, D.P., Hoppin, J.A., 2015. Pesticide exposure and end-stage renal disease risk among wives of pesticide applicators in the agricultural health study. Environ. Res. 143, 198–210.

Moss, S.R., 2010. Non-chemical methods of weed control: benefits and limitations.Seventeenth Australasian Weeds Conference, September 26, pp. 14–19.

Myers, J.P., Antoniou, M.N., Blumberg, B., Carroll, L., Colborn, T., Everett, L.G., Hansen, M., Landrigan, P.J., Lanphear, B.P., Mesnage, R., Vandenberg, L.N., 2016. Concerns over use of glyphosate-based herbicides and risks associated with exposures: a consensus statement. Environ. Health 15, 19. https://doi.org/10.1186/s12940-016-0117-0.

Nishioka, M.G., Burkholder, H.M., Brinkman, M.C., Gordon, S.M., Lewis, R.G., 1996. Measuring transport of lawn-applied herbicide acids from turf to home: correlation of dislodgeable 2,4-D turf residues with carpet dust and carpet surface residues. Environ. Sci. Technol. 30, 3313–3320.

Oerke, E.C., 2006. Crop losses to pests. J. Agric. Sci. 144, 31–43.

Pérez-Ruíz, M., Slaughter, D.C., Fathallah, F.A., Gliever, C.J., Miller, B.J., 2014. Co-robotic intra-row weed control system. Biosyst. Eng. 126, 45–55.

Peterson, G.E., 1967. The discovery and development of 2,4-D. Agric. Hist. 41 (3), 243–254.

Potts, S.G., Imperatriz-Fonseca, V., Ngo, H.T., Aizen, M.A., Biesmeijer, J.C., Breeze, T.D., Dicks, L.V., Garibaldi, L.A., Hill, R., Settele, J., Vanbergen, A.J., 2016. Safeguarding pollinators and their values to human well-being. Nature 540, 220–229.

Powles, S.B., 2008. Evolved glyphosate-resistant weeds around the world: lessons to be learnt. Pest Manag. Sci. 64 (4), 360–365.

Rohr, J.R., Palmer, B.D., 2005. Aquatic herbicide exposure increases salamander desiccation risk eight months later in a terrestrial environment. Environ. Toxicol. Chem. 24 (5), 1253–1258.

Rueda-Ayala, V., Peteinatos, G., Gerhards, R., Andújar, D., 2015. A non-chemical system for online weed control. Sensors 15 (4), 7691–7707.

Sardana, V., Mahajan, G., Jabran, K., Chauhan, B.S., 2017. Role of competition in managing weeds: an introduction to the special issue. Crop. Prot. 95, 1–7.

Shahzad, M., Farooq, M., Jabran, K., Hussain, M., 2016. Impact of different crop rotations and tillage systems on weed infestation and productivity of bread wheat. Crop. Prot. 89, 161–169.

Shaner, D.L., Beckie, H.J., 2014. The future for weed control and technology. Pest Manag. Sci. 70 (9), 1329–1339.

Sterling, T.D., Arundel, A.V., 1986. Health effects of phenoxy herbicides: a review. Scand. J. Work Environ. Health 12, 161–173.

Tadeo, J.L., Sanchez-Brunete, C., Perez, R.A., Fernández, M.D., 2000. Analysis of herbicide residues in cereals, fruits and vegetables. J. Chromatogr. A 882, 175–191.

Tshewang, S., Sindel, B.M., Ghimiray, M., Chauhan, B.S., 2016. Weed management challenges in rice (*Oryza sativa* L.) for food security in Bhutan: a review. Crop. Prot. 90, 117–124.

Ward, M.H., Lubin, J., Giglierano, J., Colt, J.S., Wolter, C., Bekiroglu, N., Camann, D., Hartge, P., Nuckols, J.R., 2006. Proximity to crops and residential exposure to agricultural herbicides in Iowa. Environ. Health Perspect. 114, 893–897.

Winston, R.L., Schwarzländer, M., Hinz, H.L., Day, M.D., Cock, M.J., Julien, M.H., 2014. Biological Control of Weeds: A World Catalogue of Agents and Their Target Weeds, fifth ed. USDA Forest Service, Forest Health Technology Enterprise Team, Morgantown, USA.

Further Reading

Hamill, A.S., Holt, J.S., Mallory-Smith, C.A., 2004. Contributions of weed science to weed control and management. Weed Technol. 18, 1563–1565.

2

Thermal Weed Control: History, Mechanisms, and Impacts

Arslan M. Peerzada, Bhagirath S. Chauhan

The University of Queensland, Gatton, QLD, Australia

2.1 INTRODUCTION

Weed control in crop production involves higher costs than disease and insect pest control because weeds are a relatively constant problem and insect and disease break out sporadically (Gianessi et al., 2003). Despite billion dollars spent on weed management, weeds cause significant losses (13.2%) in terms of reduced crop yield potential (Oerke et al., 1994). Recently, researchers have been focusing on nonherbicidal methods for controlling weeds, including physical, mechanical, and biological, as well as direct (i.e., electric charge, radiation, and flame) or indirect use of energy (i.e., solarization and mulching). From the 1940s to 1960s, before selective herbicide came in use, thermal weed control was a practical and commercial weed control options in the United States, particularly in corn and cotton (Collins, 1999). However, the developments were comparatively slower in Europe and did not fade out entirely, with the revival in use with organic growers in the 1980s.

Thermal weed control technology plays an important role in managing weeds under different agrosystems, particularly in organic agriculture. This technology is based on the principles of plant thermal energy exchange (a continuous process of energy exchange between plant and its environment) at high temperatures, which disturb the functioning of the aboveground plant parts (Sirvydas et al., 2002). Long-period exposure of plants to high temperatures up to 45 and 55°C results in plant death, which formed the basis for the development of weed management strategies involving heat or high temperature (Zimdahl, 2013). High temperatures are lethal to plant tissues as they disrupt most of the physiological functions through causing membrane rupture, protein denaturation, and enzyme deactivation (Bajwa et al., 2015). Several factors influence the heat injury on plants, including temperature, energy input, exposure period, and weed species. Most of the heat injury methods will affect the aboveground portion of the plants; however, some plants (i.e., perennial weeds) may regrow from belowground parts and thus require repeated application (Ghosh and Dolai, 2014). This chapter highlights the effects of heat on plants and environment, the various thermal weed control methods, the mechanism of action of these methods, and their advantages and disadvantages.

2.2 EFFECT OF HEAT ON PLANT AND ENVIRONMENT

2.2.1 Heat Versus Plant Growth

2.2.1.1 Seed Mortality or Reduced Vigor

High temperatures, either due to fire or heat, are reported to influence numerous aspects of seed biology, including seed viability, dormancy, and germination (Ribeiro et al., 2013). The high temperature in association with other factors will either alter the structures enclosing the embryo and prevent radicle emergence or probably alter the expression of certain temperature-regulated genes inhibitory to germination (Hills and van Staden, 2003). Inhibition of seed germination in high temperatures is also related to the induction of abscisic acid (ABA; Essemine et al., 2010). High temperatures significantly influence the establishment and survival of plant species through reduced seedling emergence, poor seedling vigor, abnormal seedlings, and reduced plumule/radicle growth (Toh et al., 2008; Piramila et al., 2012). For example, Dahlquist et al. (2007) observed seed mortality in *Sonchus oleraceus* L., *Echinochloa colona* (L.) Link, *Solanum nigrum* L., *Portulaca oleracea* L., *Sisymbrium irio* L., and *Amaranthus albus* L. at a temperature of >50°C. Similarly, temperature above 35°C has been reported to inhibit germination in the woody weed *Piper aduncum* L. (Wen et al., 2015).

2.2.1.2 Photosynthesis Inhibition

Inhibition of photosynthesis seems to be the most critical factor in heat stress (Allakhverdiev et al., 2008). It exerts a significant effect on the photosynthetic activity of C_3 plants than that of C_4 (Yamori et al., 2014). High temperatures directly influence the carbon metabolism in the stroma and phytochemical reactions in the thylakoid lamellae in the chloroplast. The major alternations in the chloroplast under heat stress include altered structural organization of thylakoids, the loss of grana stacking, and grana swelling (Ashraf and Hafeez, 2004; Rodríguez et al., 2005).

An increase in temperature above optimum results in inhibition of gross photosynthesis and ultimately speeds up the respiration and photorespiration within the affected plants. This imbalance in photosynthesis and respiration is damaging as it depletes the carbohydrate reserves rapidly. Further increase in the temperature inhibits the respiration and membrane transports, which results in plant cell death. Injury due to heat stress causes a more acute impact on the light reaction, as plants exposed for a short time cause impairment of structural organization and long-term inhibition of photosystem II (PSII) activity (Lichtenthaler et al., 2005). It damages the oxygen-evolving complex of PSII and disturbs the electron transfer within the PSII reaction (Kouril et al., 2004). In addition to this, heat stress affects the activation of ribulose-1,5-bisphosphate carboxylase (RuBisCO) due to low thermal stability of RuBisCO activase activity (DeRidder and Salvucci, 2007) and the Calvin cycle reaction, which in combination affect the leaf water status, leaf stomatal conductance, and intercellular CO_2 concentrations (Crafts-Brandner and Salvucci, 2000; Wang et al., 2010).

Heat stress affects the sucrose and starch synthesis due to reduced activity of sucrose-phosphate synthase, ADP-glucose pyrophosphorylase, and invertase (Rodríguez et al., 2005; Djanaguiraman et al., 2009). Researchers associated the photosynthesis inhibition under heat stress with reduced nitrogen (N) assimilation and decreased N availability to photosynthetic apparatus, which ultimately affects the RuBisCO protein and activity (Xu and Zhou, 2006). High temperatures also impose a negative impact on the plant leaves through reducing leaf water potential and leaf area and causing premature leaf senescence, which negatively influences the photosynthetic activity of the plants (Young et al., 2004; Greer and Weedon, 2012). For survival, plants usually accumulate proline as an adaptive mechanism for survival (Hare et al., 1998; Kavi Kishor et al., 2005), which regulates the heat stress by detoxifying reactive oxygen species (ROS), cellular osmotic adjustment, biological membrane protection, and enzyme/protein stabilization (Hare et al., 1998). The predominant pathway for proline biosynthesis under stress conditions appears to be from glutamine in plants (Delauney and Verma, 1993). Its metabolism involves pyrroline-5-carboxylate (P5C) synthase enzyme complex (γ-glutamyl kinase and γ-glutamyl phosphate reductase), regarded as the key enzymes for proline metabolism.

2.2.1.3 Metabolic Imbalance

Metabolic imbalance occurs when plants grow outside the optimum thermal range (Bita and Gerats, 2013; Martins et al., 2014). This imbalance due to thermal differences may result in a shortage of essential metabolites or buildup of toxic substances, which inhibit the normal plant functioning, that is, photosynthesis and respiration (Hasanuzzaman et al., 2013). Resultantly, the carbon fixation rate falls as temperature increases, while the carbon use may rise (referred as the thermal compensation point). Beyond this point, the plant begins to use carbohydrate reserves, and net uptake becomes negative at high temperatures. Thus, the rate of respiration falls as plants acclimate to high temperature, which decreases the impact on net photosynthesis. Metabolic imbalance can also occur due to the thermal effect on physical processes; for example, the oxygen solubility increases more than CO_2 as temperature increases, and oxygen becomes more concentrated in the plant cell as compared with CO_2. This imbalance contributes to increased RuBP oxygenation at high temperatures, resulting in increased photorespiration rates.

2.2.1.4 Membrane Permeability

The structure and fluidity of lipid membrane depend on their composition and prevailing temperature. An increase in temperature increases the lipid membrane fluidity due to weak hydrogen bonding between the adjuvant fatty acids. This increase in fluidity is linked with an uncontrolled increase in membrane permeability due to the disrupted activity of membrane-bound proteins. During exposure to high temperatures, membrane-associated processes, such as photosynthesis and membrane transport, are the first to be inhibited (Allakhverdiev et al., 2008). Similarly, high-temperature sensitivity of PSII is thought to be closely related with the thylakoid membrane. In addition, change in membrane fluidity during heat stress acts as a signal to initiate other stress responses in the plant cell (Mittler et al., 2012).

2.2.1.5 Enzyme Denaturation and Protein Turnover

Most of the enzymes are thermolabile, which lose their catalytic activity and lead toward their denaturation as the temperature increases (Salvucci and Crafts-Brandner, 2004). Also, they impair the synthesis of substitute enzymes and other cell proteins, causing overall limitation due to reduced protein turnover. Prolonged heat stress causes enzyme denaturation, which in combination with membrane dysfunctioning results in plant cell death. For example, inhibition of photosynthesis is related to the reduced catalytic activity of RuBisCO under moderate heat stress due to the thermal sensitivity of RuBisCO activase (Lazar et al., 2013). However, production of heat-stable forms of this enzyme has been observed to play an important role in high-temperature acclimation (Yamori et al., 2013).

2.2.1.6 Inhibition of Reproductive Development

Sexual reproduction in flowering plants is very sensitive to environmental stresses, particularly to thermal insults in the warm season (Giorno et al., 2013). Thermal sensitivity of reproductive processes is often the critical factor for plant productivity; a few degree elevation in temperature during flowering results in loss of plant cycle (Lobell et al., 2011). Depending on the intensity, duration, and increase rate, temperature fluctuations during flowering induce morphological changes within the flower organs (Giorno et al., 2013). Heat stress negatively affects the reproductive development by inhibiting the floral development, fertilization, and postfertilization processes in many plant species (Porter, 2005); however, thermal sensitivity at flowering varies significantly among the plant species and their ecotypes (Sato et al., 2006). Pollen viability is reported as more vulnerable to heat stress.

High temperature induces changes in the respiration and photosynthesis, which leads toward shortened life cycle and diminished plant productivity (Bita and Gerats, 2013). Possible reasons for the increased sterility under high temperature are impaired pollen germination and stunted pollen tube growth, reduced ovule viability, anomaly in stigmatic and style positions, reduced number of pollen grains retained by stigma, disturbed fertilization processes, obstacle in the endosperm growth, proembryo, and unfertilized embryo (Cao et al., 2008). Temperature-induced male sterility has been associated with morphological alternations in the sporophytic tissues, such as tapetum, epidermis, endothecium, and stomium (Sato et al., 2000, 2002). However, most of the significant changes occur in the tapetum layer and mature microspores that show alternation in the vacuolization. These tapeta serve as a nutritive source for microspore development; thus, high temperature induces tapetal defects, which strongly affects the progression of male gametogenesis and the formation of microspore cells, referred as male-sterile mutants (Scott et al., 2004; Ma, 2005).

2.2.1.7 Reduced Plant Biomass

Under moderate heat stress, inhibition of photosynthesis and other growth-related processes results in minor reductions in dry matter production (Winter and Königer, 1991). However, increase in temperature above the plant thermal ranges decreases the plant growth duration following increased growth rate. Further increase in temperature does not allow plant growth rate to compensate for the reduction in development duration and thus reduces the final plant biomass at maturity. Reduced duration of development also decreases the number of plant organs, such as seeds and fruits (Stone and Nicolas, 1994). This type of plant response can easily be observed in plant organs, such as leaves, stems, and fruits. At maturity, small organ size under high temperature is associated with smaller cells rather than cell numbers; it indicates that cell enlargement is more sensitive to high temperature than cell division. Under certain conditions, plants accumulate sugar in their leaves due to altered sucrose-metabolizing enzyme activity when grown under heat stress (Jie et al., 2012), which indicates that translocation can be more limiting than photosynthesis.

2.2.2 Effects of Heat on Soil Properties

During heating, the spatial distribution of soil properties within a soil profile determines the magnitude of changes occurring in a particular property to a large extent (DeBano, 1991). For example, properties of soil located near or on the surface are more likely to be altered due to direct exposure to surface heating. In this regard, organic matter and related soil properties are more likely to be modified than other properties (i.e., clay content) by radiated energy, and it is directly related to the change in soil chemical properties.

2.2.2.1 *Physical Properties*

High temperature alters the soil physical properties, which create potential risks of subsidence, erosion, and other environmental hazards (Zihms et al., 2013). Most of the soil physical properties related to soil organic matter (except clay content), such as soil structure, pore space, and soil aggregation, are affected by extreme heat. Arocena and Opio (2003) observed that increased soil temperature induces dehydration of 2:1 clay minerals leading toward a strong clay particle interaction, which produces less clay and more silt-sized particles in the soil. In addition to this, high temperature creates heat-induced cracks in the sand-sized particles that result in the breakdown and reduces sand-sized particles in the soil (Pardini et al., 2004). It also lowers the clay content and increases the silt content (Inbar et al., 2014). Increases in soil temperature over 30°C increase the aggregate stability of the soil due to the thermal transformation of iron aluminum oxides, which act as cementing agents for clay particles to form strong silt-sized particles in the soil (Fox et al., 2007). Increased soil temperature decreases the water viscosity and reduces the soil moisture content, allowing more water to percolate through the soil profile (Broadbent, 2015). Increased soil temperature maximizes the evaporation rate and restricts the water movement into the soil profile. High temperature encourages the soil microbial activity, which alters the soil aeration through increasing the CO_2 content in the soil (Allison, 2005).

2.2.2.2 *Chemical Properties*

High temperature decreases the soil organic matter through combustion, which also reduces the clay size fraction, leading toward reduced cation-exchange capacity in the soil (Ubeda et al., 2009). Denaturation of soil organic matter at high temperature increases the soil pH (Menzies and Gillman, 2003). Nutrients contained in the fossil fuels and soil organic matter are usually cycled in the environment by biological decomposition processes when the temperature reaches 38°C, and sufficient soil moisture is available to sustain microbial activity. Under normal conditions, soil microorganisms decompose organic matter and slowly release nutrients; however, these nutrients undergo an irreversible transformation in the case of heating. The response of individual nutrients differs and has its inherent temperature threshold, which is divided into three categories: sensitive, moderately sensitive, and relatively insensitive. Depending upon this, nitrogen and sulfur are considered as sensitive, potassium and phosphorus as moderately sensitive, and calcium, magnesium, and manganese as relatively insensitive. Nutrients with low threshold temperatures are easily subjected to volatilization, thus resulting in nutrient losses. Despite this, high temperature also influences the nutrient availability to plants either by in situ changes or by translocation of organic substance downward into the soil.

2.2.2.3 *Biological Properties*

Soil temperature (10–28°C) significantly affects the soil respiration through increasing the extracellular enzyme activity, which degrades the soil polymeric organic matter and increases the microbial soluble substance retake and thus increases the microbial respiration rate (Conant et al., 2008; Allison et al., 2010). Soil nitrogen mineralization increases with increased soil temperature due to amplified microbial activity and organic matter decomposition. The sensitivity of soil organic matter decomposition to temperature change is also critical to the global carbon balance (Pare et al., 2006). At soil temperature between 21 and 28°C, organic matter decomposition increases due to increase in the movement of the soluble substrate in the soil and stimulating microbial activity (Fierer et al., 2005; Broadbent, 2015).

Soil microorganisms are probably more sensitive to high temperature because they are living organisms that have relatively low lethal temperature thresholds. Temperature is reported as an important element, which influences the biochemical and physical processes of microorganisms, their activity, and population in the soil (Bagyaraj and Rangaswami, 2007). Microorganisms can survive in extreme temperature conditions (-60 and $+60$°C); however, the optimum temperature range is rather narrow at which soil microbes grow and function actively. In the geothermal field on Earth, microorganisms are diverse and abundant with some microbes that are capable of surviving at high temperatures of up to 122°C (Cavicchioli, 2002; Stieglmeier et al., 2014). During extreme temperature conditions, microbial activity at the topsoil is supposed to be highly reduced (Conant et al., 2011).

Temperature influences the soil biota activity directly through affecting the physiological activities (i.e., enzyme activity) or indirectly through altering physicochemical properties, such as nutrient diffusion and solubility, mineral weathering, and evaporation rate (Paul, 2006; A'Bear et al., 2014). On the basis of temperature ranges at which soil microbes grow and function, they are divided into three types, that is, psychrophiles, mesophilic, and thermophilic, which grow at <10, 20–45, and >45°C, respectively. Generally, mesophilic bacteria under high temperature undergo a declined activity and survival period although other microorganisms adapted to these temperatures may succeed in finding alternative temporal conditions to develop, may be due to extreme enzymatic activity (Gonzalez et al., 2015). In some cases, potential activity of mesophilic bacteria has been observed at a temperature above 40°C or after exposure at this temperature due to metabolic inhibition or germination of resting cells (Ho and Frenzel 2012; Gonzalez et al., 2015).

2.3 THERMAL WEED CONTROL TECHNOLOGIES

Stubble burning, a traditional way of thermal weed control, has been effectively used for reducing the number of viable weed seeds returned to the soil after crop harvest; this technique has been banned due to smoke and other fire-related hazards. Thermal weed control involves heat energy to kill emerged weeds and their seed production through the involvement of different direct and indirect techniques. Currently, different techniques or methods of thermal weed control have been involved using various energy sources to generate heat for killing weed seeds and seedlings. These techniques usually include soil solarization, use of infrared and microwave radiations, lasers, electrostatic fields, irradiation, electrocution, electrostatic fields, ultraviolet lights, flame, steam, and hot water (Heisel et al., 2002; Mathiassen et al., 2006; Ascard et al., 2007; Sivesind et al., 2009).

2.3.1 Soil Solarization

Soil solarization, known as soil sterilization or pasteurization, has been successfully used as a non-chemical method to control or reduce soilborne insect pests, pathogens, mites, and weeds. This technique has been reported as one of the first thermal weed control methods, which heats soil by using plastic mulches. Polythene sheets trap the heat from the solar radiation and increase it to the level that is lethal to the weed seeds and seedlings (Abouziena and Haggag, 2016). It requires prolonged exposure to sunlight with increased ambient temperatures for a sufficient period to reduce the weed seed populations in the soil (Golzardi et al., 2015; Mauro et al., 2015). In addition, this technique improves crop germination through attaining optimal soil temperature and prevents the establishment of soilborne diseases and insects through sterilizing heat effect. This technique has been extensively used for high-value horticultural crops, such as lettuce, garlic, tomato, and squash, which can compensate the increased costs of solarization. Soil temperature, mulching material, soil moisture, and climatic conditions are considered as critical factors in affecting the solarization efficiency (Dai et al., 2016).

Over the past three decades, soil solarization has been evaluated as a non-chemical and sustainable method for controlling soilborne pathogens and weeds in many countries (McGovern et al., 2013). In modern agriculture, transparent and black polyethene plastic sheets have been used for the pre- and postplanting control of weed species, respectively. Polyethene, polyvinyl chloride, and ethylene-vinyl acetate are among the most common plastic films used in agriculture, which perform much better than black sheets in soil solarization (Reddy, 2013). Due to the flexibility, tensile strength, and puncture resistance, low-density polyethene has been widely used as agricultural mulch. The use of double-layer polyethene sheets is the improved soil solarization technology that causes a 3–10°C increase in soil temperatures as compared with single one (Barakat and Al-Masri, 2012). Air layer between these two layers acts as an insulation against the escape of both heat and moisture from the soil.

2.3.1.1 Mechanism

It is a hydrothermal disinfection method in which moist soil is covered usually using a colored polyethene sheet during the hot summer months. This process causes in thermal, physical, biological, and chemical changes within the soil (Stapleton and DeVay, 1986). The principle behind this technique is the increase in temperature up to the lethal level in the top soil where most of the dormant and viable seeds are located (Soumya et al., 2004). Solar radiations are composed of shortwave solar radiations and longwave terrestrial radiations. The shortwave radiations pass through the polyethene sheet, and longwave terrestrial radiations are held back, which results in increasing soil temperature, lethal to the insect pests, pathogens, and weeds. From the moistened soil, the evaporating water will condense on the inner side of the sheet and will drop back to the soil surface. These condense droplets prevent the escape of longwave radiations and thus minimize the cooling process of the soil surface. In order to increase the efficiency of this method, necessary measures should be taken to prevent heat losses through sensible and latent heat fluxes; therefore, the mulch should be intact with the soil surface (Mahrer et al., 1984).

Soil solarization has been attributed to reducing the pest intensity through the direct or indirect mode of actions. In the direct pest inactivation, the competitive ability of the unwanted plant is suppressed by plant exposure to lethal temperatures (McGovern and McSorley, 1997). However, the indirect effect reduces the pest densities through increased antagonistic microorganism population, which are thermophilic and can recolonize in solarized soil. It also alters the physical and chemical properties of the soil, resulting in an increased soil moisture level and volatile accumulation; alters the soil-gas composition that is lethal to the pests; and alters the nutrient availability to the plants (Patricio et al., 2006). This technique is usually employed for one month or longer duration during the summers, particularly in areas with little cloud cover and rainfall (Katan, 1980). In some cases, integration of soil solarization with other control practices (biological organisms, soil amendment, etc.) may enhance the effectiveness of this technique (McGovern et al., 2013).

2.3.1.2 Effect on Weed Growth and Reproduction

Soil solarization reduces the intensity of all weed types: narrow-leaved, sedges, and broad-leaved (Marenco and Lustoca, 2000). However, some weeds are moderately resistant, requiring optimum conditions, such as good soil moisture, tight-fitting plastic, and high radiations for control (Elmore, 1995). Soil solarization helps in depleting the soil reserves of dormant weed seeds, which acts as a primary source of persistent weed problems (Ghosh and Dolai, 2014). The possible effects of soil solarization on the weeds include breaking weed seed dormancy, solar scorching of newly emerged seedlings, direct killing by heat, and indirect microbial killing of weed seeds, which are weakened by heating (Ghosh and Dolai, 2014). In addition, it also affects the germination of weed seeds, particularly buried near the soil surface; soil temperature increases at the surface and decreases with the increase in the soil depth; thus, weed seeds buried deep in the soil may survive (Lalitha et al., 2003).

Research reported that weed species like *Chamaecrista nictitans* var. *paraguariensis* (Chod & Hassl.) Irwin & Barneby, *Marsypianthes chamaedrys* (Vahl) O. Kuntze, *Mitracarpus* spp., *Mollugo verticillata* L., *Sebastiania corniculata* M. Arg., and *Spigelia anthelmia* L. can effectively be suppressed by using transparent polyethene sheets (Marenco and Lustoca, 2000). However, most of the difficult-to-control perennial weed species, like *Cyperus* spp., *Sorghum halepense* (L.) Pers., *Convolvulus arvensis* L., and *Cynodon dactylon* (L.) Pers. having buried reproductive organs (i.e., rhizomes, tubers, and stolen), affect the practicability of soil solarization (DeVay et al., 1991; Marenco and Lustoca, 2000). A significant difference has been observed due to solarization duration and its interaction with the cultivated soil depth as it works more effectively in tilled soils. Soil solarization integrated with green manuring and poultry manure suppressed the emergence of *Poa annua* L. and *Cuscuta campestris* Yuncker (Haidar and Sidahmed, 2000; Peachey et al., 2001). Studies showed that the success of this non-chemical weed management option usually depends on the light exposure, soil texture, soil moisture, and weed flora. Integration of soil solarization with cultivation might be a reliable option for controlling persistent weed species, including the perennial yield affecters (e.g., *S. halepense*, *Cyperus rotundus*, and *C. dactylon*).

2.3.1.3 Merits and Demerits

Soil solarization is an eco-friendly method that is compatible with organic and integrated crop management systems. It is efficient in controlling soilborne pathogens with a long-term effect on an agroecosystem, regarding soil nutrient improvement and microbial activities. Soil solarization increased the growth response and harvestable yield with no hazards to the growers, workers, or public. However, it requires a crop-free field for a relatively long period (1–2 months) and also requires supplement irrigation. Also, there are chances of weed seed survival when buried deep in the soil. Soil solarization may have some environmental pollution issues from the plastic disposal after the treatment (Dai et al., 2016).

2.3.2 Flaming

Flaming is the most common method of thermal weeding in which intense heat wave is used to rupture the plant cells (Bond and Grundy, 2001). In this technique, weeds are usually killed by heat in the form of fire. To create intense flame, liquefied petroleum gas (LPG) or propane is used as a fuel source (Ascard, 1995). However, the fuel and temperature requirements depend upon the weed growth stage and biomass. Most commonly, propane has been used as a fuel in different flame weeders; however, hydrogen is supposed to be a relatively renewable alternative (Andersen 1997). These systems have been developed from handheld flamers to tractor mounted systems to be used in small-scale vegetable production to large-scale row-cropping systems, respectively (Bond and Grundy, 2001). Indifferent cropping systems, flaming provided effective weed control and helped in maintaining the stability of the system. However, its utilization as an effective tool in reducing weed competition and recycling soil nutrient is confined to areas where the fire is not a threat (Kyser and DiTomaso, 2002). For example, flame methods are not considered as a reliable option in countries like Australia, as these provide unacceptable fire risks and are unsuitable for urban conditions.

2.3.2.1 History

In organic cropping systems, flaming is considered as one of the most promising options for weed control, and it can also be used in conventional crops (Knezevic and Ulloa, 2007). For the first time, this tool was used in 1852 by John A. Craig of Arkansas, who developed a machine for selective flame weeding in sugarcane. In the mid-1940s, when kerosene and oil were replaced by more efficient LPGs, extensive use of flame weeders began (Edwards, 1964). Early systems of flame weeding were relatively crude and dangerous. With the passage of time, improvements in machinery and application protocols for selective flame weeding were developed for agronomic and horticultural crops (i.e., cotton, soybean, sorghum, potatoes, and tomatoes) (Thompson et al., 1967). In the United States, >15,000 thermal units were in use for designing flame weeders in different row crops in 1964 (Ascard, 1995). In the late 1950s, the popularity of these techniques started to decline due to increase in petroleum prices and with the availability of efficient and less expensive herbicides (Daar and Lennox, 1987). Flaming equipment with a range of burners has been developed in several countries, including Germany, the United Kingdom, Holland, Sweden, and Denmark (Holmoy and Netland, 1994; Caspell, 2002).

2.3.2.2 Mechanism

In this technique, flaming kills the plant tissues by exposing these to an intense heat wave; it does not ignite the plant tissues, a misconception (Leroux et al., 2001). Thermal energy applied to plant in combination with exposure time determines the effectiveness of this technique in controlling weeds (Ascard, 1995; Seifert and Snipes, 1996). Previous study showed that temperature ranges from 50 to 95°C with an exposure time of 0.65–0.13 s that is necessary to kill weeds (Daniell et al., 1969). Burner generates heat waves with a combustion temperature of 1900°C, which rapidly increases the temperature in the exposed plant tissues. LPG is commonly used to feed burners and generates flames with an average temperature of 1500°C (Ascard, 1995).

Direct heat injury to targeted plants results in membrane protein denaturation, enzyme deactivation, and loss in cell functioning, resulting in plant death or reduce its competitive ability drastically. Fingerprint test can be used to access the effectiveness of the flaming treatment by pressing the leaf between the thumb and index finger (Knezevic et al., 2012). After pressing the leaf surface firmly, if darkened impression is visible, it means the leaf losses the internal pressure due to water leakage from the ruptured cell walls. Numerous factors positively or negatively affect the effectiveness of this weed control (Mojžiš and Varga, 2013). For example, weed types, their growth stages, prevailing weather conditions, crop type, and heat absorption and transmission influence the precise setting of the flame weeder. Therefore, variability within the flaming process should be eliminated to reduce the negative impacts on achieving the best results against weed species. Mojžiš and Varga (2013) recommended that operation speed and gas pressure of the flame weeder needs to be adjusted according to the weed species, their growth stage, and the crop when treated to increase the efficacy of thermal weed control.

2.3.2.3 Effect on Weed Growth and Reproduction

Researchers reported flaming for its best control of broad-leaved weeds, including *Abutilon theophrasti* Medik. and *Amaranthus retroflexus* L., whereas the results were not acceptable for grassy weeds, for example, *Setaria viridis* (L.) Beauv. and *Echinochloa crus-galli* (L.) P. Beauv (Knezevic and Ulloa, 2007). However, growth stage at the time of flaming determines the sensitivity of weeds to heat (Martelloni et al., 2016). Most of the researchers believe that flaming at early growth stage should be recommended for effective weed control (Sivesind et al., 2009; Ulloa et al., 2010; Knezevic et al. 2014). Rask et al. (2013) reported some positive results regarding the control of grass weeds.

In European countries, the flame weeders are used where weeds with thin-layered leaves can easily be burnt with a single operation, whereas weeds like *Capsella bursa-pastoris* (L.) Medic., *E. colona*, and *P. annua* require more applications (Ascard 1995). Flame weeding prior to crop emergence showed positive results on newly emerged weeds in potato, sugar beet, carrots, and cayenne pepper (Melander, 1998). In lowbush blueberry (*Vaccinium angustifolium* Ait.) fields, weed seeds exposed to direct flaming for 1 s reduced the germination of *Tragopogon pratensis* L. Tropr., *Apocynum androsaemifolium* L. Apcan., and *Panicum capillare*

L. Panca (White and Boyd, 2016). In a recent study, it has been observed that two cross flaming applied separately with an LPG dose (36–42 kg ha^{-1}) provided adequate control of *Chenopodium album* L., *Datura stramonium* L., *A. retroflexus*, and *C. dactylon* with economically acceptable yield in maize (Martelloni et al., 2016).

2.3.2.4 Merits and Demerits

If we compare flaming with herbicide application, this technique definitely has few advantages. For example, there are no residual impact on plants, soil, air, or water with no drift hazards and herbicide carryover to the next season and can control herbicide-resistant weeds more effectively (Wszelaki et al., 2007). In addition, it also provides benefits over manual or mechanical cultivation in terms of minimum labor requirement, reduced costs, and less soil disturbance (which prevent weed seed transfer to the upper soil surface and reduce soil erosion) and provides control of pathogens and insects (Lague et al., 1997).

Despite all these benefits, flame weeding also has some disadvantages when compared with other weed management options. In comparison with herbicides, flame weeders are costly as compared with herbicide applicators. Furthermore, it has low application speed, the lack of crop selectivity, low field capacity, and the lack of residual weed control (Ascard, 1995; Ascard et al., 2007). Most of the commonly used flame weeders have the same field capacity as mechanical cultivators for weed control (Ascard et al., 2007). In addition, due to the combustion of propane gas or diesel fuel, this flaming process releases greenhouse gas CO_2 in the atmosphere. Ulloa et al. (2011) estimated that propane at 60 kg ha^{-1} produced 188.9 kg CO_2 ha^{-1} from propane and diesel combustion. This high energy requirement and release of CO_2 emission could be seen as a disadvantage.

2.3.3 Hot Water

Heat can be used to kill weeds through the application of hot water; a popular technique to kill weeds in cracks and driveways. It is considered as one of the safest, simple, and economical technique for thermal weed control with no harmful effects, like flaming and microwave radiations. It has been considered as an effective method to control many annual weeds and also restrict the growth of perennial weeds. The effectiveness of this technique is high under dense weed populations due to increased penetration ability (Hansson and Ascard, 2002). In European countries, hot water application is considered in precision weed management strategies due to its greater success rate.

2.3.3.1 History

In the early 1900s, hot water application for the control of weeds has been investigated with the great zeal of success in different countries. The introduction of this weed control techniques eliminated the fire hazards and energy consumption issues associated with flame weeding (Hansson and Ascard, 2002). Aqua-Hot, a commercial equipment, was developed in the United States for weed control, which proved effective against several annual and perennial weed species, and was comparable with the results of glyphosate (Berling, 1992; Rask and Kristoffersen, 2007). Later on, a similar equipment called "the Waipuna system" was used in New Zealand in landscape and vegetation management, requiring single or multiple hot water treatment for the control of annual and perennial weeds on long-term basis; the system was specialized with hot foam to achieve an insulating effect and to keep hot water in contact with weed species for a long time (Quarles, 2004). Nowadays, hot water equipment "H$_2$O Hot Aqua Weeder" is available in Denmark and the Netherlands, particularly for horticultural crops. Kempenaar and Spijker (2004) stated that addition of foam cover increased the efficiency of this system; however, decreasing the operation speed did not have any significant impact. Compared with flaming, the ability to penetrate into the vegetation is greater for hot water. On the other hand, the energy efficiency is relatively high, and water applicators need to be improved to decrease heat losses (Hansson, 2002). In this regard, patch sprayer equipped with digital cameras helped in saving energy and a further increase in treatment capacity (Hansson and Ascard, 2002). A few years back, a system with precision application, known as "wave machine," has been development in the Netherlands.

2.3.3.2 Mechanism

Hot water melts the membrane cuticle of the plant leaves and breaks down the plant cell structures, which makes the plant unable to retain moisture and dehydrate within few hours or days. This technology has been equally effective for controlling germinating and mature plants. Factors like weed development stage, exposure time, droplets size, wetting agent, water temperature, daytime variation in thermal sensitivity, and water flow were observed to influence the hot water treatment for controlling weeds (Hansson and Mattsson, 2002; De Cauwer et al., 2015). In addition to this, environmental factors like air temperature, drought, and rainfall are also hypothesized to affect the hot water weed control efficacy (Hansson and Mattsson, 2003).

2.3.3.3 Effect on Weed Growth and Reproduction

Studies showed that hot water could be used for controlling annual and young perennial weeds, and their results were comparable with glyphosate treatment. Most of the weed species regrow after non-chemical treatment, requiring consecutive treatments (Melander et al., 2009). Therefore, some hot water applications (mostly ranging from 3 to 5) may be required during the growing season to keep weeds under control (Rask et al., 2013). In grasses, erect growth habitat hampers the water retention on plants and makes it difficult to control by hot water (De Cauwer et al., 2015). Hot water treatment at energy dose of 589 kJ m^{-2} was highly efficient in controlling *Lolium perenne* L., *Festuca rubra* L., *Taraxacum officinale* F.H. Wigg., and *Plantago major* L. (De Cauwer et al., 2016). Due to erectophile growth, most grassy weeds are less sensitive to hot water as compared with planophile leaved weeds, that is, *T. officinale*. In organic farming, about 80% mortality rate was observed when *Rumex obtusifolius* L. was treated with hot water (90°C; Latsch and Sauter, 2014).

2.3.3.4 Merits and Demerits

In comparison with chemical weed control, hot water treatment offers multiple benefits as it does not contaminate the underground water, soil, and air and eliminates the risks of potential exposure of human/wildlife to pesticide residues. On the other hand, hot water has a broad-spectrum biocide like many chemical herbicides that also affects the beneficial soil organisms. For a large-scale application, hot water treatment is not cost-effective; therefore, some believe that the use of propane flamers or infrared heat for direct weed control is more energy-efficient and preferable over hot water.

2.3.4 Saturated Steam

High-temperature water steam as thermal weed control technology is promising for the timely suppression of weed population. Weed control through steam is also observed as a more effective, quick, and sustainable method to control weeds on relatively hard surfaces (Rask and Kristoffersen, 2007). Over the past few years, this technology has been used for weed control in areas where complete vegetation removal is required. It can also be used as a spot treatment for controlling weeds in turf or cracks.

2.3.4.1 History

Previously, steaming was used to sterile the soil and control both weeds and diseases before the crop establishment under laboratory conditions (Sonneveld, 1979; Runia, 1983). Concerns about the use of methyl bromide renewed researchers' interest in steam sterilization methods. In the early 1990s, the potential of steam to control weeds was demonstrated; however, its application as an effective weed control option was not realized (Upadhyaya et al., 1993). Later on, it was effectively used for controlling weeds in forestry and in cropping systems (Kolberg and Wiles, 2002). Different mobile steaming equipments have been available to control weeds and pathogens in polytunnels and fields.

2.3.4.2 Mechanism

In this method, a mixture of vaporized water and steam is sprayed on weeds (Wei et al., 2010). Steam possesses high energy density with a high heat transferring capability. Through gas technology, heat intensity increases 1000–2000 times when compared with flaming. Therefore, wet steam immediately increases the temperature in the plant surface tissues and causes a destructive impact. In open fields and greenhouse horticulture, various soil steaming methods have been developed, such as sheet steaming and steaming with fixed pipes, which are costly and laborious. The work efficiency and fuel consumption of these methods were greatly influenced by the weed groups and their growth stages, soil volume and texture, exposure time, and temperature (Barberi et al., 2009; Peruzzi et al., 2012). In the recent years, efforts have been made to improve the steam supply method to reduce work capacity (Vidotto et al., 2013). These improvements resulted in the development of self-propelled machines (i.e., band-steaming machine and Ecostar SC600 with blade) and adding active compounds, such as KOH and CaO, to increase soil temperature (Barberi et al., 2009; Peruzzi et al., 2012). In an open field, different environmental factors, such as air temperature, soil moisture, and soil texture, are likely to influence the susceptibility of seeds to steam heating (Gay et al. 2010; Melander and Kristensen, 2011). This highlights the importance to bring the information on various environmental factors on steam method under consideration.

2.3.4.3 Effect on Weed Growth and Reproduction

Single application of steam can eliminate most of the annual weeds and early perennial weeds, whereas mature perennial weeds require two applications for acceptable control (Wei et al., 2010). In addition, this method has less effect on the underground plant portion, but repeated application kills the aboveground plant part(s) and thus restricts the plant nutrition, which ultimately results in plant death (Rifai et al., 2002; Wei et al., 2010). Short time exposure for 2–4 s using superheated steam (400°C) killed the seedlings of numerous weed species (Upadhyaya et al., 1993). It was observed that extent of heat injury depends on the weed species, growth stage, steam temperature, and exposure duration. Weed exposure to a very short duration (0.1–0.2 s) was found effective in damaging the foliar epicuticular wax, affected the cell membrane integrity, and reduced the seed production. It also helped in killing weed seeds; imbibed seeds were generally more susceptible. However, in some weed species, seed coats may offer protective covering from the steam. Most of the perennial weed species regenerate, making the repeated exposure necessary.

Kolberg and Wiles (2002) reported that steam application at 3200 L ha^{-1} provided 90% control of C. album and A. retroflexus; results were almost similar to glyphosate application (560 g ai/ha^{-1}). However, the emergence of C. album, A. retroflexus, and S. nigrum was not affected by the steam application. In the case of weed biomass, reduced values were observed at 9 weeks after treatment in all the tested weeds when treated at the seedling stage, whereas Bromus tectorum L. biomass was reduced when steam was applied at anthesis. Steaming for 3 min at low temperatures (50–60°C) followed by an 8 min resting period in the steamed soil and immediate removal from the soil thereafter resulted in 100% control of C. album and Elymus repens L. (van Loenen et al., 2003). Melander and Jørgensen (2005) observed that rise in soil temperature by about 70°C reduced the seedling emergence of weed species by 99%. Similarly, band steaming with a maximum temperature of 86°C at 40 mm depth reduced 90% of the annual weed population in the field (Hansson and Svensson, 2007). Barberi et al. (2009) determined the weed suppression potential of soil steaming plus activating compound (Cao and KoH) to boost soil temperature. Activated soil steaming clearly suppressed the weed seed bank of Alopecurus myosuroides Huds., Matricaria chamomilla L., Raphanus raphanistrum L., A. retroflexus, E. colona, Fallopia convolvulus (L.) A. Love, and S. viridis (L.) P. Beauv.; however, steaming alone suppressed the emergence of A. myosuroides by 77%. Surface steaming at 0–7 cm depth was most effective that controlled weeds up to 100%. However, deep steaming (i.e., 7–14 and 14–21 cm depth) depleted the weed seed bank up to 95% in deep soil layers (Peruzzi et al., 2012).

2.3.4.4 Merits and Demerits

In hot steam weeding, heat can be directed onto unwanted vegetation, more precisely. In addition, hot moisture-laden air sinks into targeted plants rather than rising (like in flaming) with no fire risks. It has less effect on crop plants and beneficial insects. As compared with hot water treatment, it uses less water and provides better canopy penetration. Steam can be superheated to a very high temperature to increase the efficacy and shortens the exposure time duration. It also has high heat transmission coefficient as compared with hot water, ensuring more heat transmission to the plant during contact (Wei et al., 2010). On the other hand, steam is easier to volatilize than water, so heat loses very quickly. The major drawback of this method is its costs as it has a high consumption of fossil fuels and is laborious. The initial investment in steam equipment is very high, and environment and weather conditions seriously affect the operational activity (Riley, 1995). Although the steam application has been used both in open fields and greenhouses, it has limited application in protecting high-value crops. In addition to this, it is difficult to apply this technique on broad scale for the protection of major agronomic crops, that is, rice, soybean, and wheat. Therefore, application of steam in the broad area requires technological innovations allowing fast operation and lower costs.

2.3.5 Hot Foam

Hot foam weed control method is a nontoxic technique and is applicable for numerous weed species (Wei et al., 2010). In the emerging markets of organic farming and sustainable agricultural production, this method has diverse practical values and importance in the future. Despite weed control in cropped areas, hot foam has been used to manage weeds in public areas and hard surfaces to reduce the possibilities of water and air contamination in urban areas. Instead of this, thermal foam has been diversely used to control mold, bacteria, and other serious pests in the soil (Rajamannan, 1996; Beilfuss et al., 2009).

2.3.5.1 History

Heated foam for weed control was first patented in 1995 and further developed by Weeding Technologies Ltd., GB (Rajamannan, 1996). The author hypothesized that vegetation treated with hot water would die as a result of heated liquid that melts the cuticle wax of the plant. However, temperature surrounding the vegetation must be maintained for long duration for the effectiveness of the process. Later, Tindall et al. (2002) and James et al. (2009) presented the method and apparatus for weed control with hot foam.

2.3.5.2 Mechanism

In this method, a superheated air is compressed and used to melt the plant's waxy cuticle while using foam in this process as an insulating agent with a temperature close to the boiling point of water and surfactant combination (Rajamannan, 1996). The melting of the waxy cuticle causes dehydration and ultimately the plant death. Foam films prevent heat from rapid releasing with the hot water flowing, ensuring relatively stable temperature around the weed. This foam can be derived from the mixture of coconut sugar and corn sugar (Shu-ren et al., 2007). In Sweden, alkyl polyglucosides have been used as a foaming agent with hot water to be used for controlling weeds in rail yards and railways (Cederlund and Börjesson, 2016).

2.3.5.3 Effect on Weed Growth and Reproduction

Due to its antisag property, hot foam keeps in contact with weeds for a long time and on large surface area, thus increasing the heat exchange time and area, which is beneficial for managing weeds with high stalked (Wei et al., 2010; Cederlund and Börjesson, 2016). In addition, this method has the advantage of less weather change susceptibility, security, high application accuracy, high speed, and low cost (Wei et al., 2010). Spraying biodegradable Waipuna hot foam three times at a 1 month interval killed 50% of the *Crassula helmsii* (Kirk) Cockayne, which was similar to the results obtained from glyphosate (Bridge, 2005). Limited information is available regarding the influence of hot foam on the growth, development, and seed production of weed species yet.

2.3.5.4 Merits and Demerits

Hot foam has been considered as the most efficient thermal weed control option as compared with hot air, open flame, or steam (Rask et al., 2013). It has an advantage over hot water as it requires less water and has more heat transfer time (James et al., 2009). This long heat transfer period will help in providing selective weed control and economy by using low temperatures. It does not have any impact on the soil properties and can also be correlated with mechanical cultivation.

2.4 CONCLUSIONS

This chapter highlighted the history, mechanisms, and impact use of different thermal weed control technologies under field conditions. It not only described the environmental benefits in terms of soil and water quality but also discussed the negative impact on air quality and energy usage. It is suggested that thermal weed control methods may be integrated with other non-chemical weed control techniques to achieve integrated weed control. Furthermore, improvements should be made in existing thermal weed control options to increase the energy use efficiency and practicability of these systems in different farming systems.

References

A'Bear, A.D., Jones, T.H., Kandeler, E., Boddy, L., 2014. Interactive effects of temperature and soil moisture on fungal-mediated wood decomposition and extracellular enzyme activity. Soil Biol. Biochem. 70, 151–158.
Abouziena, H.F., Haggag, W.M., 2016. Weed control in clean agriculture: a review. Planta Daninha 34, 377–392.
Allakhverdiev, S.I., Kreslavski, V.D., Klimov, V.V., Los, D.A., Carpentier, R., Mohanty, P., 2008. Heat stress: an overview of molecular responses in photosynthesis. Photosynth. Res. 98, 541–550.
Allison, S.D., 2005. Cheaters, diffusion and nutrients constrain decomposition by microbial enzymes in spatially structured environments. Ecol. Lett. 8, 626–635.
Allison, S.D., Wallenstein, M.D., Bradford, M.A., 2010. Soil-carbon response to warming dependent on microbial physiology. Nat. Geosci. 3, 336–340.
Andersen, J., 1997. Experimental trials and modelling of hydrogen and propane burners for use in selective flaming. Biol. Agric. Hortic. 14, 207–219.
Arocena, J.M., Opio, C., 2003. Prescribed fire-induced changes in properties of sub-boreal forest soils. Geoderma 113, 1–16.
Ascard, J., 1995. Effects of flame weeding in weed species at different developmental states. Weed Res. 35, 397–411.
Ascard, J., Hatcher, P.E., Melander, B., Upadhyaya, M.K., 2007. Thermal weed control. In: Upadhyaya, M.K., Blackshaw, R.E. (Eds.), Non-Chemical Weed Management: Principles, Concepts and Technology. In: vol. 1. CAB International, Oxfordshire, UK, pp. 155–175.
Ashraf, M., Hafeez, M., 2004. Thermotolerance of pearl millet and maize at early growth stages: growth and nutrient relations. Biol. Plant. 48, 81–86.
Bagyaraj, D.J., Rangaswami, G., 2007. Agricultural Microbiology. PHI Learning Pvt. Ltd, New Delhi, India.
Bajwa, A.A., Mahajan, G., Chauhan, B.S., 2015. Nonconventional weed management strategies for modern agriculture. Weed Sci. 63, 723–747.

Barakat, R.M., AL-Masri, M.I., 2012. Enhanced soil solarization against *Fusarium oxysporum* f. sp. *lycopersici* in the uplands. Int. J. Agric. Res. 2012, 368654.

Barberi, P., Moonen, A.C., Peruzzi, A., Fontanelli, F., Raffaelli, M., 2009. Weed suppression by soil steaming in combination with activating compounds. Weed Res. 49, 55–66.

Beilfuss, W., Gradtke, R., Goroncy-Bermes, P., Behrends, S., Mohr, M., 2009. U.S. Patent No. 7,481,973. U.S. Patent and Trademark Office. Washington, DC, USA.

Berling, J., 1992. Getting weeds in hot water. A new hot water weed sprayer and soy-based oil help cut herbicide use. Farm Indus. News 26, 44–49.

Bita, C., Gerats, T., 2013. Plant tolerance to high temperature in a changing environment: scientific fundamentals and production of heat stress-tolerant crops. Front. Plant Sci. 4, 273.

Bond, W., Grundy, A.C., 2001. Non-chemical weed management in organic farming systems. Weed Res. 41, 383–405.

Bridge, T., 2005. Controlling New Zealand pygmyweed *Crassula helmsii* using hot foam, herbicide and by burying at Old Moor RSPB Reserve, South Yorkshire, England. Conserv. Evid. 2, 33–34.

Broadbent, F.E., 2015. Soil organic matter. Sustain. Options Land Manage. 2, 34–38.

Cao, Y.Y., Duan, H., Yang, L.N., Wang, Z.Q., Zhou, S.C., Yang, J.C., 2008. Effect of heat stress during meiosis on grain yield of rice cultivars differing in heat tolerance and its physiological mechanism. Acta Agron. Sin. 34, 2134–2142.

Caspell, N., 2002. Goodness Gracious Great Balls of Fire!! The Vegetable Farmer, Act Publishing, Maidstone, UK.

Cavicchioli, R., 2002. Extremophiles and the search for extraterrestrial life. Astrobiology 2, 281–292.

Cederlund, H., Börjesson, E., 2016. Hot foam for weed control—do alkyl polyglucoside surfactants used as foaming agents affect the mobility of organic contaminants in soil? J. Hazard. Mater. 314, 312–317.

Collins, M., 1999. In: Bishop, A.C., Boersma, M., Barnes, C.D. (Eds.), Thermal weed control, a technology with a future.12th Australian Weeds Conference-Weed Management into the 21st Century: Do We Know Where we're Going. Wrest Point Convention Centre, Hobart, Tasmania.

Conant, R.T., Drijber, R.A., Haddix, M.L., Parton, W.J., Paul, E.A., Plante, A.F., Steinweg, J.M., 2008. Sensitivity of organic matter decomposition to warming varies with its quality. Glob. Chang. Biol. 14, 868–877.

Conant, R.T., Ryan, M.G., Ågren, G.I., Birge, H.E., Davidson, E.A., Eliasson, P.E., Evans, S.E., Frey, S.D., Giardina, C.P., Hopkins, F.M., Hyvönen, R., 2011. Temperature and soil organic matter decomposition rates—synthesis of current knowledge and a way forward. Glob. Chang. Biol. 17, 3392–3404.

Crafts-Brandner, S.J., Salvucci, M.E., 2000. Rubisco activase constrains the photosynthetic potential of leaves at high temperature and CO_2. Proc. Natl. Acad. Sci. 97, 13430–13435.

Daar, A.S., Lennox, E.S., 1987. Tumor markers and antigens. In: Daar, A.S., Woodruf, M. (Eds.), Tumor Markers in Clinical Practice. In: vol. 1. Blackwell Scientific Publications, Oxford, UK, pp. 1–26.

Dahlquist, R.M., Prather, T.S., Stapleton, J.J., 2007. Time and temperature requirements for weed seed thermal death. Weed Sci. 55, 619–625.

Dai, Y., Senge, M., Yoshiyama, K., Zhang, P., Zhang, F., 2016. Influencing factors, effects and development prospect of soil solarization. Rev. Agric. Sci. 4, 21–35.

Daniell, J.W., Chappell, W.E., Couch, H.B., 1969. Effect of sublethal and lethal temperature on plant cells. Plant Physiol. 44, 1684–1689.

De Cauwer, B., Bogaert, S., Claerhout, S., Bulcke, R., Reheul, D., 2015. Efficacy and reduced fuel use for hot water weed control on pavements. Weed Res. 55, 195–205.

De Cauwer, B., De Keyser, A., Biesemans, N., Claerhout, S., Reheul, D., 2016. Impact of wetting agents, time of day and periodic energy dosing strategy on the efficacy of hot water for weed control. Weed Res. 56, 323–334.

DeBano, L.F., 1991. In: Harvey, A.E., Neuenschwander, L.F. (Eds.), The effect of fire on soil properties. Proceedings Management and Productivity of Western-Montane Forest Soils. U.S. Department of Agriculture, Forest Service, Intermountain Research Station, Ogden, UT, pp. 151–155.

Delauney, A.J., Verma, D.P.S., 1993. Proline biosynthesis and osmoregulation in plants. Plant J. 4, 215–223.

DeRidder, B.P., Salvucci, M.E., 2007. Modulation of Rubisco activase gene expression during heat stress in cotton (*Gossypium hirsutum* L.) involves post-transcriptional mechanisms. Plant Sci. 172, 246–254.

DeVay, J., Stapleton, J., Elmore, C., 1991. Soil Solarization. FAO Plant Production and Protection Paper 109. Food and Agriculture Organization, Italy, Rome.

Djanaguiraman, M., Sheeba, J.A., Devi, D.D., Bangarusamy, U., 2009. Cotton leaf senescence can be delayed by nitrophenolate spray through enhanced antioxidant defense system. J. Agron. Crop Sci. 195, 213–224.

Edwards, F.E., 1964. In: History and progress of flame cultivation. Proceedings of the First Annual Symposium of Research on Flame Weed Control. Memphis, Tennessee, USA, pp. 3–6.

Elmore, C., 1995. Soil solarization, a non pesticidal method for controlling diseases, nematodes and weeds. In: Taller Regional de Solarizacion del Suelo. Food and Agriculture Organization, Italy, Rome, pp. 18–21.

Essemine, J., Ammar, S., Bouzid, S., 2010. Impact of heat stress on germination and growth in higher plants: physiological, biochemical and molecular repercussions and mechanisms of defence. J. Biol. Sci. 10, 565–572.

Fierer, N., Craine, J.M., McLauchlan, K., Schimel, J.P., 2005. Litter quality and the temperature sensitivity of decomposition. Ecology 86, 320–326.

Fox, D.M., Darboux, F., Carrega, P., 2007. Effects of fire-induced water repellency on soil aggregate stability, splash erosion, and saturated hydraulic conductivity for different size fractions. Hydrol. Process. 21, 2377–2384.

Gay, P., Piccarolo, P., Aimonino, D.R., Tortia, C., 2010. A high efficiency steam soil disinfestation system, part I: physical background and steam supply optimization. Biosyst. Eng. 107, 74–85.

Ghosh, P., Dolai, A.K., 2014. Soil solarization, an eco-physiological method of weed control. Global J. Sci. Front. Res. 14, 43–44.

Gianessi, L., Sankula, S., Reigner, N., 2003. Plant Biotechnology: Potential Impact for Improving Pest Management in European Agriculture. A Summary of Nine Case Studies. The National Center for Food and Agricultural Policy, Washington, DC.

Giorno, F., Wolters-Arts, M., Mariani, C., Rieu, I., 2013. Ensuring reproduction at high temperatures: the heat stress response during anther and pollen development. Plants 2, 489–506.

Golzardi, F., Vaziritabar, Y., Vaziritabar, Y., Asilan, K.S., Sayadi, M.H.J., Sarvaramini, S., 2015. Effect of solarization and polyethylene thickness cover type on weeds seed Bank and soil properties. JAEBS 5, 88–95.

Gonzalez, J.M., Portillo, M.C., Piñeiro-Vidal, M., 2015. Latitude-dependent underestimation of microbial extracellular enzyme activity in soils. Int. J. Environ. Sci. Technol. 12, 2427–2434.

Greer, D.H., Weedon, M.M., 2012. Modelling photosynthetic responses to temperature of grapevine (*Vitis vinifera* cv. Semillon) leaves on vines grown in a hot climate. Plant Cell Environ. 35, 1050–1064.

Haidar, M.A., Sidahmed, M.M., 2000. Soil solarization and chicken manure for the control of *Orobanche crenata* and other weeds in Lebanon. Crop. Prot. 19, 169–173.

Hansson, D., 2002. Hot Water Weed Control on Hard Surface Areas (Ph.D. thesis). Department of Agricultural Engineering, Swedish University of Agricultural Sciences. Alnarp, Sweden. Report 323.

Hansson, D., Ascard, J., 2002. Influence of developmental stage and time of assessment on hot water weed control. Weed Res. 42, 307–316.

Hansson, D., Mattsson, J.E., 2002. Effect of drop size, water flow, wetting agent and water temperature on hot-water weed control. Crop. Prot. 21, 773–781.

Hansson, D., Mattsson, J.E., 2003. Effect of air temperature, rain and drought on hot water weed control. Weed Res. 43, 245–251.

Hansson, D., Svensson, S.-E., 2007. In: Steaming soil in narrow bands to control weeds in row crops. Proceedings of the 7th EWRS (European Weed Research Society) Workshop on Physical and Cultural Weed Control, Salem, Germany.

Hare, P.D., Cress, W.A., van Staden, J., 1998. Dissecting the roles of compatible osmolyte accumulation during stress. Plant Cell Environ. 21, 535–553.

Hasanuzzaman, M., Nahar, K., Fujita, M., 2013. Extreme Temperature Responses, Oxidative Stress and Antioxidant Defense in Plants. INTECH Open Access Publisher, Rijeka, Croatia.

Heisel, T., Schou, J., Andreasen, C., Christensen, S., 2002. Using laser to measure stem thickness and cut weed stems. Weed Res. 42, 242–248.

Hills, P.N., Van Staden, J., 2003. Thermoinhibition of seed germination. S. Afr. J. Bot. 69, 455–461.

Ho, A., Frenzel, P., 2012. Heat stress and methane-oxidizing bacteria: effects on activity and population dynamics. Soil Biol. Biochem. 50, 22–25.

Holmoy, R., Netland, J., 1994. Band spraying, selective flame weeding and hoeing in late white cabbage—Part I. Acta Hort. 372, 223–234.

Inbar, A., Lado, M., Sternberg, M., Tenau, H., Ben-Hur, M., 2014. Forest fire effects on soil chemical and physicochemical properties, infiltration, runoff, and erosion in a semiarid Mediterranean region. Geoderma 221, 131–138.

James, G.W., Reid, D.G., Tindall, D.W., 2009. Method for Weed Control With Hot Foam, US. Patent No. 20050005509.

Jie, Z., Xiaodong, J., Tianlai, L., Zaiqiang, Y., 2012. Effect of moderately-high temperature stress on photosynthesis and carbohydrate metabolism in tomato (*Lycopersico esculentum* L.) leaves. Afr. J. Agric. Res. 7 (3), 487–492.

Katan, J., 1980. Solar pasteurization of soils for disease control: status and prospects. Plant Dis. 64, 450–454.

Kavi Kishor, P.B., Sangam, S., Amruth, R.N., Sri Laxmi, P., Naidu, K.R., Rao, K.R.S.S., Reddy, K.J.S.R., Theriappan, P., Sreenivasulu, N., 2005. Regulation of proline biosynthesis, degradation, uptake and transport in higher plants: its implications in plant growth and abiotic stress tolerance. Curr. Sci. 88, 424–438.

Kempenaar, C., Spijker, J.H., 2004. Weed control on hard surfaces in The Netherlands. Pest Manag. Sci. 60, 595–599.

Knezevic, S.Z., Datta, A., Bruening, C., Gogos, G., 2012. Propane-Fueled Flame Weeding in Corn, Soybean, and Sunflower. University of Nebraska—Lincoln and PERC extension publication, Washington DC, USA.

Knezevic, S.Z., Stepanovic, S., Datta, A., 2014. Growth stage affects response of selected weed species to flaming. Weed Technol. 28, 233–242.

Knezevic, S., Ulloa, S., 2007. Potential new tool for weed control in organically grown agronomic crops. J. Agric. Sci. 52, 95–104.

Kolberg, R.L., Wiles, L.J., 2002. Effect of steam application on cropland weeds. Weed Technol. 16, 43–49.

Kouril, R., Lazár, D., Ilík, P.P., Skotnica, J., Krchnák, P., Naus, J., Krcha'k, P., 2004. High temperature induced chlorophyll fluorescence rise in plants at 40–50°C: experimental and theoretical approach. Photosynth. Res. 81, 49–66.

Kyser, G.B., DiTomaso, J.M., 2002. Instability in a grassland community after the control of yellow starthistle (*Centaurea solstitialis*) with prescribed burning. Weed Sci. 50, 648–657.

Lague, C., Gill, J., Lehoux, N., Peloquin, G., 1997. Engineering performances of propane flamers used for weed, insect pest, and plant disease control. Appl. Eng. Agric. 13, 7–16.

Lalitha, B.S., Nanjappa, H.V., Ramachandrappa, B.K., 2003. Effect of soil solarization on soil microbial population and the germination of weed seeds in the soil. J. Ecol. 15, 169–173.

Latsch, R., Sauter, J., 2014. Optimisation of hot-water application technology for the control of broad-leaved dock (*Rumex obtusifolius*). J. Agric. Eng. Res. 45, 137–145.

Lazar, D., Murch, S.J., Beilby, M.J., Al Khazaaly, S., 2013. Exogenous melatonin affects photosynthesis in characeae *Chara australis*. Plant Signal. Behav. 8, e23279.

Leroux, G.D., Douhéret, J., Lanouette, M., 2001. Flame weeding in corn. In: Vincent, C., Panneton, B., Fleurat-Lessard, F. (Eds.), Physical Control Methods in Plant Protection. In: vol. 1. Springer-Verlag, Berlin Heidelberg, UK, pp. 47–60.

Lichtenthaler, H.K., Langsdorf, G., Lenk, S., Buschamann, C., 2005. Chlorophyll fluorescence imaging of photosynthetic activity with the flesh lamp fluorescence imaging system. Photosynthetica 43, 355–369.

Lobell, D.B., Schlenker, W., Costa-Roberts, J., 2011. Climate trends and global crop production since 1980. Science 333, 616–620.

Ma, H., 2005. Molecular genetic analyses of microsporogenesis and microgametogenesis in flowering plants. Annu. Rev. Plant Biol. 56, 393–434.

Mahrer, Y., Naot, O., Rawitz, E., Katan, J., 1984. Temperature and moisture regimes in soils mulched with transparent polyethylene. Soil Sci. Soc. Am. J. 48, 362–367.

Marenco, R.A., Lustosa, D.C., 2000. Soil solarization for weed control in carrot. Pesq. Agrop. Brasileira 35, 2025–2032.

Martelloni, L., Fontanelli, M., Frasconi, C., Raffaelli, M., Peruzzi, A., 2016. Cross-flaming application for intra-row weed control in maize. Appl. Eng. Agric. 32, 569–578.

Martins, L.D., Tomaz, M.A., Lidon, F.C., DaMatta, F.M., Ramalho, J.C., 2014. Combined effects of elevated [CO_2] and high temperature on leaf mineral balance in *Coffea* spp. plants. Clim. Chang. 126, 365–379.

Mathiassen, S.K., Bak, T., Christensen, S., Kudsk, P., 2006. The effect of laser treatment as a weed control method. Biosyst. Eng. 95, 497–505.

Mauro, R.P., Monaco, A.L., Lombardo, S., Restuccia, A., Mauromicale, G., 2015. Eradication of *Orobanche/Phelipanche* spp. seedbank by soil solarization and organic supplementation. Sci. Hortic. 193, 62–68.

McGovern, R.J., Chaleeprom, W., Chaleeprom, W., McGovern, P., To-Anun, C., 2013. Evaluation of soil solarization and amendments as production practices for lettuce and vegetable soybean in northern Thailand. J. Agric. Technol. 9, 1863–1872.

McGovern, R.J., McSorley, R., 1997. Physical methods of soil sterilization for disease management including soil solarization. In: Rechcigl, N.A., Rechcigl, J.A. (Eds.), Environmentally Safe Approaches to Crop Disease Control. CRC Press, Boca Raton, FL, pp. 283–313.

Melander, B., 1998. Interaction between soil cultivation in darkness, flaming, and brush weeding when used for in-row weed control in vegetables. Biol. Agric. Hortic. 16, 1–14.

Melander, B., Holst, N., Grundy, A.C., Kempenaar, C., Riemens, M.M., Verschwele, A., Hansson, D., 2009. Weed occurrence on pavements in five North European towns. Weed Res. 49, 516–525.

Melander, B., Jørgensen, M.H., 2005. Soil steaming to reduce intrarow weed seedling emergence. Weed Res. 45, 202–211.

Melander, B., Kristensen, J.K., 2011. Soil steaming effects on weed seedling emergence under the influence of soil type, soil moisture, soil structure and heat duration. Ann. Appl. Biol. 158, 194–203.

Menzies, N.W., Gillman, G.P., 2003. Plant growth limitation and nutrient loss following piled burning in slash and burn agriculture. Nutr. Cycl. Agroecosyst. 65, 23–33.

Mittler, R., Finka, A., Goloubinoff, P., 2012. How do plants feel the heat? Trends Biochem. Sci. 37, 118–125.

Mojžiš, M., Varga, F., 2013. Effect of setting the parameters of flame weeder on weed control effectiveness. J. Cent. Eur. Agric. 14, 1373–1380.

Oerke, E.C., Dehene, H., Schoenbeck, F., Weber, A., 1994. Rice losses. In: Amsterdam, B.V. (Ed.), Crop Production and Crop Protection, Estimated Losses in Major Food and Cash Crops. Elsevier Science, Amsterdam, The Netherland.

Pardini, G., Gispert, M., Dunjó, G., 2004. Relative influence of wildfire on soil properties and erosion processes in different Mediterranean environments in NE Spain. Sci. Total Environ. 328, 237–246.

Pare, D., Boutin, R., Larocque, G.R., Raulier, F., 2006. Effect of temperature on soil organic matter decomposition in three forest biomes of eastern Canada. Can. J. Soil Sci. 86, 247–256.

Patricio, F.R.A., Sinigaglia, C., Barros, B.C., Freitas, S.S., Neto, J.T., Cantarella, H., Ghini, R., 2006. Solarization and fungicides for the control of drop, bottom rot and weeds in lettuce. Crop. Prot. 25, 31–38.

Paul, E.A., 2006. Soil Microbiology, Ecology and Biochemistry. Academic Press, Burlington, USA.

Peachey, R.E., Pinkerton, J.N., Ivors, K.L., Miller, M.L., Moore, L.W., 2001. Effect of soil solarization, cover crops, and metham on field emergence and survival of buried annual bluegrass (Poa annua) seeds. Weed Technol. 15, 81–88.

Peruzzi, A., Raffaelli, M., Frasconi, C., Fontanelli, M., Bàrberi, P., 2012. Influence of an injection system on the effect of activated soil steaming on Brassica juncea and the natural weed seedbank. Weed Res. 52, 140–152.

Piramila, B.H.M., Prabha, A.L., Nandagopalan, V., Stanley, A.L., 2012. Effect of heat treatment on germination, seedling growth and some biochemical parameters of dry seeds of black gram. IJPPR 1, 194–202.

Porter, J.R., 2005. Rising temperatures are likely to reduce crop yields. Nature 436, 174.

Quarles, W., 2004. Thermal weed management: hot alternatives for urban areas and organic farms. IPM Pract. 26, 1–9.

Rajamannan, A.H.J., 1996. U.S. Patent No. 5,575,111. Washington, DC: U.S. Patent and Trademark Office.

Rask, A.M., Kristoffersen, P., 2007. A review of non-chemical weed control on hard surfaces. Weed Res. 47, 370–380.

Rask, A.M., Larsen, S.U., Andreasen, C., Kristoffersen, P., 2013. Determining treatment frequency for controlling weeds on traffic islands using chemical and non-chemical weed control. Weed Res. 53, 249–258.

Reddy, P.P., 2013. Soil solarization. In: Reddy, P.P. (Ed.), Recent Advances in Crop Protection. Springer, India; Heidelberg, pp. 159–184.

Ribeiro, L.C., Pedrosa, M., Borghetti, F., 2013. Heat shock effects on seed germination of five Brazilian savanna species. Plant Biol. 15, 152–157.

Rifai, M.N., Astatkie, T., Lacko-Bartosova, M., Gadus, J., 2002. Effect of two different thermal units and three types of mulch on weeds in apple orchards. J. Environ. Eng. Sci. 1, 331–338.

Riley, B., 1995. Hot water: a "cool" new weed control method. J. Pestic. Reform 15, 9.

Rodríguez, M., Canales, E., Borrás-Hidalgo, O., 2005. Molecular aspects of abiotic stress in plants. Biotecnol. Apl. 22, 1–10.

Runia, W.T., 1983. A recent development in steam sterilization. Acta Hortic. 152, 195–200.

Salvucci, M.E., Crafts-Brandner, S.J., 2004. Inhibition of photosynthesis by heat stress: the activation state of Rubisco as a limiting factor in photosynthesis. Physiol. Plant. 120, 179–186.

Sato, S., Kamiyama, M., Iwata, T., Makita, N., Furukawa, H., Ikeda, H., 2006. Moderate increase of mean daily temperature adversely affects fruit set of Lycopersicon esculentum by disrupting specific physiological processes in male reproductive development. Ann. Bot. 97, 731–738.

Sato, S., Peet, M.M., Thomas, J.F., 2000. Physiological factors limit fruit set of tomato (Lycopersicon esculentum Mill) under chronic, mild heat stress. Plant Cell Environ. 23, 719–726.

Sato, S., Peet, M.M., Thomas, J.F., 2002. Determining critical pre- and post-anthesis periods and physiological processes in Lycopersicon esculentum Mill exposed to moderately elevated temperatures. J. Exp. Bot. 53, 1187–1195.

Scott, R.J., Spielman, M., Dickinson, H.G., 2004. Stamen structure and function. Plant Cell 16, S46–S60.

Seifert, S., Snipes, C.E., 1996. Influence of flame cultivation on mortality of cotton (*Gossypium hirsutum*) pests and beneficial insects. Weed Technol. 10, 544–549.

Shu-ren, C., Li, Y., Pan, L., 2007. Review and prospect of thermal weed control technologies. J. Anhui Agric. Sci. 35, 10695–10697.

Sirvydas, A., Lazauskas, P., Vasinauskienơ, R., Kerpauskas, P., 2002. In: Thermal weed control by water steam. 5th EWRS Workshop on Physical Weed Control, Pisa, Italy, pp. 253–262.

Sivesind, E.C., Leblanc, M.L., Cloutier, D.C., Seguin, P., Stewart, K.A., 2009. Weed response to flame weeding at different developmental stages. Weed Technol. 23, 438–443.

Sonneveld, C., 1979. Changes in chemical properties of soil caused by steam sterilization. In: Mulder, D. (Ed.), Developments in Agricultural and Managed-Forest Ecology 6: Soil Disinfestation. Elsevier, New York, pp. 39–49.

Soumya, T.M., Nanjappa, H.V., Ramachandrappa, B.K., 2004. Effect of soil solarization on weed count, weed dry weight and pod yield of groundnut. J. Agric. Sci. 17, 548–550.

Stapleton, J.J., De Vay, J.E., 1986. Soil solarization: a non-chemical method for management of plant pathogens and pests. Crop. Prot. 5, 190–198.

Stieglmeier, M., Klingl, A., Alves, R.J., Simon, K.M.R., Melcher, M., Leisch, N., Schleper, C., 2014. *Nitrososphaera viennensis* gen. nov., sp. nov., an aerobic and mesophilic, ammonia-oxidizing archaeon from soil and a member of the archaeal phylum Thaumarchaeota. Int. J. Syst. Evol. Microbiol. 64, 2738–2752.

Stone, P.J., Nicolas, M.E., 1994. Wheat cultivars vary widely in their responses of grain yield and quality to short periods of post-anthesis heat stress. Funct. Plant Biol. 21, 887–900.

Thompson, V.J., Scheibner, R.A., Thompson, W.C., 1967. In: Flaming of alfalfa in Kentucky for weevil and weed control—Results of 1967 winter. American Society of Agricultural Engineers. St. Joseph, Ml, USA, ASAE Paper, pp. 637–645.

Tindall, D., David, R., Gibb, J., 2002. Method and Apparatus for Weed Control with Hot Foam. U.S. Patent Application 10/491, 312.

Toh, S., Imamura, A., Watanabe, A., Nakabayashi, K., Okamoto, M., Jikumaru, Y., Hanada, A., Aso, Y., Ishiyama, K., Tamura, N., Iuchi, S., 2008. High temperature-induced abscisic acid biosynthesis and its role in the inhibition of gibberellin action in Arabidopsis seeds. Plant Physiol. 146, 1368–1385.

Ubeda, X., Pereira, P., Outeiro, L., Martin, D.A., 2009. Effects of fire temperature on the physical and chemical characteristics of the ash from two plots of cork oak (*Quercus suber*). Land Degrad. Dev. 20, 589–608.

Ulloa, S.M., Datta, A., Malidza, G., Leskovsek, R., Knezevic, S.Z., 2010. Timing and propane dose of broadcast flaming to control weed population influenced yield of sweet maize (*Zea mays* L. var. *rugosa*). Field Crop Res. 118, 282–288.

Ulloa, S.M., Datta, A., Bruening, C., Neilson, B., Miller, J., Gogos, G., Knezevic, S.Z., 2011. Maize response to broadcast flaming at different growth stages: effects on growth, yield and yield components. Eur. J. Agron. 34, 10–19.

Upadhyaya, M.K., Polster, D.F., Klassen, M.J., 1993. Weed control by superheated steam. Weed Sci. Soc. Am. Abstr. 33, 115.

van Loenen, M.C., Turbett, Y., Mullins, C.E., Feilden, N.E., Wilson, M.J., Leifert, C., Seel, W.E., 2003. Low temperature-short duration steaming of soil kills soil-borne pathogens, nematode pests and weeds. Eur. J. Plant Pathol. 109, 993–1002.

Vidotto, F., De Palo, F., Ferrero, A., 2013. Effect of short-duration high temperatures on weed seed germination. Ann. Appl. Biol. 163, 454–465.

Wang, L.J., Fan, L., Loescher, W., Duan, W., Liu, G.J., Cheng, J.S., Luo, H.B., Li, S.H., 2010. Salicylic acid alleviates decreases in photosynthesis under heat stress and accelerates recovery in grapevine leaves. BMC Plant Biol. 10, 34.

Wei, D., Liping, C., Zhijun, M., Guangwei, W., Ruirui, Z., 2010. Review of non-chemical weed management for green agriculture. Int. J. Agric. Biol. Eng. 3, 52–60.

Wen, B., Xue, P., Zhang, N., Yan, Q., Ji, M., 2015. Seed germination of the invasive species piper aduncum as influenced by high temperature and water stress. Weed Res. 55, 155–162.

White, S.N., Boyd, N.S., 2016. Effect of dry heat, direct flame, and straw burning on seed germination of weed species found in lowbush blueberry fields. Weed Technol. 30, 263–270.

Winter, K., Königer, M., 1991. Dry matter production and photosynthetic capacity in *Gossypium hirsutum* L. under conditions of slightly suboptimum leaf temperatures and high levels of irradiance. Oecologia 87, 190–197.

Wszelaki, A.L., Doohan, D.J., Alexandrou, A., 2007. Weed control and crop quality in cabbage (*Brassica oleracea* (capitata group)) and tomato (*Lycopersicon lycopersicum*) using a propane flamer. Crop. Prot. 26, 134–144.

Xu, Z.Z., Zhou, G.S., 2006. Nitrogen metabolism and photosynthesis in *Leymus chinensis* in response to long-term soil drought. J. Plant Growth Regul. 25, 252–266.

Yamori, W., 2013. Improving photosynthesis to increase food and fuel production by biotechnological strategies in crops. J. Plant Biochem. Physiol. 1, 113.

Yamori, W., Hikosaka, K., Way, D.A., 2014. Temperature response of photosynthesis in C3, C4, and CAM plants: temperature acclimation and temperature adaptation. Photosynth. Res. 119, 101–117.

Young, L.W., Wilen, R.W., Bonham-Smith, P.C., 2004. High temperature stress of *Brassica napus* during flowering reduces micro- and megagametophyte fertility, induces fruit abortion, and disrupts seed production. J. Exp. Bot. 55, 485–495.

Zihms, S.G., Switzer, C., Irvine, J., Karstunen, M., 2013. Effects of high temperature processes on physical properties of silica sand. Eng. Geol. 164, 139–145.

Zimdahl, R.L., 2013. Fundamentals of Weed Science. Academic Press, San Diego.

Further Reading

Andreasen, C., Hansen, L., Streibig, J.C., 1999. The effect of ultraviolet radiation on the fresh weight of some weeds and crops. Weed Technol. 13, 554–560.

Ark, P.A., Parry, W., 1940. Application of high-frequency electrostatic fields in agriculture. Q. Rev. Biol. 15, 172–191.

Ascard, J., 1998. Comparison of flaming and infrared radiation techniques for thermal weed control. Weed Res. 38, 69–76.

Baranski, S., Czerski, P., 1976. Biological Effects of Microwaves. Dowden, Hutinson and Ross, Inc., Stroudsburg, PA.

Bayramian, A., Fay, P.E., Dyer, W.E., 1993. In: Weed control using carbon dioxide lasers. Proceedings Western Society of Weed Science, Logan, Utah, USA, vol. 45, pp. 55–56.

Biradar, I.B., Hosamani, M.M., 1997. Effect of soil solarization on weed control and its after—effects on growth and yield of groundnut (*Arachis hypogaea* L.). J. Agric. Sci. 10, 966–970.

Bloembergen, N., 1974. Laser-induced electric breakdown in solids. IEEE J. Quantum Electron. QE-10, 375–386.

Brighenti, A.M., Brighenti, D.M., 2009. Weed control in organic soybean using electrical discharge. Cienc. Rural 39, 2315–2319.

Brodie, G., 2016. In: Microwave weed control.IMPI'S 50th Annual Microwave Power Symposium, Orlando, Florida, USA.

Brodie, G., Hamilton, S., Woodworth, J., 2007. An assessment of microwave soil pasteurization for killing seeds and weeds. Plant Prot. Q. 22, 143–149.

Brodie, G., Ryan, C., Lancaster, C., 2012. Microwave technologies as part of an integrated weed management strategy: a review. Int. J. Agric. Res. 2012, 636905.

Couch, R., Gangstad, E.O., 1974. Response of waterhyacinth to laser radiation. Weed Sci. 22, 450–453.

Davis, F.S., Wayland, J.R., Merkle, M.G., 1971. Ultrahigh-frequency electromagnetic fields for weed control: phytotoxicity and selectivity. Science 173, 535–537.

Davis, F.S., Wayland, J.R., Merkle, M.G., 1973. Phytotoxicity of a UHF electromagnetic field. Nature 241, 291–292.

Day, T.A., Martin, G., Vogelmann, T.C., 1993. Penetration of UV-B radiation in foliage: evidence that the epidermis behaves as a non-uniform filter. Plant Cell Environ. 16, 735–741.

Diprose, M.E., Benson, E.A., 1984. Electrical methods of killing plants. J. Agric. Eng. Res. 30, 197–209.

Diprose, M.F., Benson, F.A., Hackam, R., 1980. Electrothermal control of weed beet and bolting sugar beet. Weed Res. 20, 311–322.

Drolet, C., Rioux, R., 1983. Evaluation d'une rampe utili sant un courant electrique pour le contröle des mauvaises herbes, ERDAF Rep. No. 345Z.0 1843-I-EC24. Res. Branch, Agric. Can., Ottawa. p. 66.

Dykes, W.G., 1980. In: Principles and practices of electrical weed control in row crops.American Society of Agriculture Engineering, St. Joseph, MI. Paper No. 80-1007, p. 6.

Feller, U., Crafts-Brandner, S.J., Salvucci, M.E., 1998. Moderately high temperatures inhibit ribulose-1,5-bisphosphate carboxylase/oxygenase (Rubisco) activase-mediated activation of Rubisco. Plant Physiol. 116, 539–546.

Flint, D.R., 1977. The toxicological significance of large animal metabolism studies with pesticides. In: Ivie, G.W., Dorough, H.W. (Eds.), Fate of Pesticides in Large Animals. Academic Press, New York.

Hamid, M.A.K., Boulanger, R.J., Tong, S.C., Gallop, R.A., Pereira, R.R., 1969. Microwave pasteurization of raw milk. J. Microw. Power 4, 272–275.

Heald, C.M., Menges, R.M., Wayland, J.R., 1974. Efficacy of ultra-high frequency electromagnetic energy and soil fumigation on the control of the reniform nematodes and common purslane among southern peas. Plant Dis. 58, 985–987.

Heisel, T., Schou, J., Christensen, S., Andreasen, C., 2001. Cutting weeds with a CO_2 laser. Weed Res. 41, 19–29.

Holmoy, R., Storeheier, K.J., 1993. In: Selective flaming in the plant row and basic investigation and development of flamers.Communications of the 4th International Conference I.F.O.A.M. Non Chemical Weed Control, Dijon, France.

Ingle, A., 1992. Performance of the Non-toxic Vegetation Control Process. Crop and Food Confidential Report No. 3. New Zealand Institute for Crop a Food Research, ltd. Levin, New Zealand.

Katan, J., 1981. Solar heating (solarization) of soil for control of soil-borne pests. Annu. Rev. Phytopathol. 19, 211–236.

Katan, J., Greenberger, A., Alon, H., Grinstein, A., 1976. Solar heating by polyethylene mulching for the control of diseases caused by soilborne pathogens. Phytopathology 66, 683–688.

Lai, R., 1974. Soil temperature, soil moisture, and maize yield from mulched and unmulched tropical soils. Plant Soil 40, 129–143.

Langerholc, J., 1979. Moving phase transition in laser irradiated biological tissue. Appl. Opt. 18, 2286–2293.

Loux, M.M., Diigab, D., Dobbles, A.F., Reeb, B., Johnson, W.G., Legleiter, T.R., 2013. Weed Control Guide for Ohio and Indiana. Ohio State University Extention and Purdue Extention, Ohio State.

Oberholtzer, L., Dimitri, C., Greene, C., 2005. Price Premiums Hold on as Organic Produce Market Expands. Outlook Report VGS-308-01 USDA Economic Research Service: 22.

Ooi, M.K.J., 2012. Seed bank persistence and climate change. Seed Sci. Res. 22, S53–S60.

Reifschneider, D., Nunn, R.C., 1965. In: Infrared cotton defoliation or desiccation. Proceedings of the Second Annual Symposium, Use of Flame in Agriculture, St. Louis, Missouri, USA, pp. 25–29.

Sanwald, E., Koch, W., 1978. In: In physical methods of weed control. Proceedings of the Brighton Crop Protection Conference on Weeds. 3, pp. 977–986.

Sartorato, I., Zanin, G., Baldoin, C., De Zanche, C., 2006. Observations on the potential of microwaves for weed control. Weed Res. 46, 1–9.

Scott, R.A., Jr., 1970. Laser Plant Control. U. S. Patent 3652844, 9. 2.

Scott, N.W., Nathan, S.B., 2016. Effect of dry heat, direct flame, and straw burning on seed germination of weed species found in lowbush blueberry fields. Weed Technol. 30, 263–270.

Unger, P.W., 1978. Straw mulch effects on soil temperatures and sorghum germination and growth. Agron. J. 70, X58–X64.

Vela, G.R., Wu, J.F., 1979. Mechanism of lethal action of 2450-MHz radiation on microorganisms. Appl. Environ. Microbiol. 37, 550–553.

Vela, G.R., Wu, J.F., Smith, D., 1976. Effect of 2450 MHz microwave radiation on some soil microorganism in situ. Soil Sci. 121, 44–51.

Vela-Múzquiz, R., 1983. In: Vanachter, A. (Ed.), Control of field weeds by microwave radiation.II International Symposium on Soil Disinfestation, Leuven, Belgium, pp. 201–208.

Vigneault, C., 2002. Weed electrocution. In: David, P. (Ed.), Encyclopedia of Pest Management. In: vol. 1. Marcel Dekker Inc., Basel, NY, pp. 896–898.

Vigneault, C., Benoit, D.L., 2001. Electrical weed control: theory and applications. In: Vincent, C., Panneton, B., Fleurat-Lessard, F. (Eds.), Physical Control Methods in Plant Protection. Springer-Verlag, Berlin, Heidelberg, UK, pp. 174–188.

Vigneault, C., Benoit, D.L., McLaughlin, N.B., 1990. Energy aspect of weed electrocution. Rev. Weed Sci. 5, 15–26.

Wayland, J.R., Merkel, M.G., Menges, R.M.m and Wilkerson, R.E., 1980. Disruption of cell membranes by electromagnetic radiation. Stud. Nat. Sci. 2, 1–6.

3

The Use of Physics in Weed Control

Graham Brodie
Melbourne University, Dookie, VIC, Australia

3.1 INTRODUCTION

Modern no-till cropping depends on herbicides for weed management; therefore, herbicide applications are an important system input. Unfortunately, herbicide resistance in many weed species is becoming wide spread (Heap, 1997, 2008), and multiple herbicide resistances in several economically important weed species have also been widely reported (Owen et al., 2007). In time, herbicide-resistant weeds may ultimately result in significant yield reductions and grain contamination.

The International Agency for Research on Cancer (IARC), which is part of the World Health Organization (WHO), has also concluded that glyphosate is probably carcinogenic to humans (Guyton et al., 2015). This announcement has generated considerable debate in the media concerning the use of herbicides. Other authors have also highlighted the potential hazard to human health of long-term exposure to herbicides and pesticides (Hernández et al., 2013; Duke, 2010; Mačkić and Ahmetović, 2011; Peighambarzadeh et al., 2011; Troudi et al., 2012; Wickerham et al., 2012); therefore, there has been growing interest in nonherbicidal control of weeds. The objectives of this chapter are to outline some of the potential technologies, apart from herbicide application, for weed management. These technologies include flaming, steam treatment, electrocution, applying electrostatic fields, microwave weed treatment, applying infrared radiation, applying ultraviolet radiation, using lasers, robotics, and using abrasive weed control techniques.

3.2 THERMAL WEED CONTROL

Thermodynamics predicts that energy, in the form of heat, moves along the temperature gradient until all spatial coordinates reach equilibrium. By definition, equilibrium is reached when the temperature gradient disappears from the system.

Heat is transferred by conduction, convection, and radiation. Conduction is the transfer of heat between solid/solid interfaces and within solids. Convection is the transfer of heat between an object and its environment due to fluid motion, that is, a gas or liquid interface with a solid. Radiation is the transfer of heat between bodies through the emission and absorption of electromagnetic energy, without the need for a fluid interface (i.e., as a purely spatial phenomenon). This section will explore how heat can be used to kill weeds.

3.2.1 Flaming

Flame weeding is the most commonly applied thermal weed control method. Several kinds of equipment have been developed for weeding, such as tractor-mounted flamers and hand-pushed or handheld devices for weeding around obstacles and for private households. Flaming controls a wide range of weed species (Ascard, 1994), some of which are tolerant or resistant toward herbicides. Flaming gave 72% and 80% control of common rye and volunteer alfalfa, respectively. Both kochia (*Kochia scoparia* (L.) Roth) and netseed lambsquarter (*Chenopodium berlandieri*) were also controlled at 65% (Gourd, 2002).

Ascard (1994) developed dose-response relationships between applied energy and weed response. He used three models to determine responses; however, his results are mostly based on the model presented in Eq. (3.1):

$$y = \frac{D}{1 + \left(\frac{x}{a}\right)^b} \tag{3.1}$$

where y is the response variable of the plant fresh weight or plant number; x is the liquid petroleum gas (LPG) consumption in $kg\,ha^{-1}$; and D, a, and b are parameters to be determined experimentally. From this model, an LD_{50} and LD_{95} were derived for white mustard (*Sinapis alba* L.) at different plant sizes and densities (Table 3.1).

Weed flamers should be shielded, preferably with a long and relatively low roofed shield (Storeheier, 1994) to keep combustion gases close to the ground for as long as possible; the burner angle should be 22.5–45 degrees to the horizontal.

Tandem burners did not increase effective ground speed compared with single burners (Ascard, 1998). According to Ascard (1994), propane doses of $10–40\,kg\,ha^{-1}$ were required to achieve 95% control of sensitive species with 0–4 leaves, while plants with 4–12 leaves required $40–150\,kg\,ha^{-1}$. At $49\,MJ\,kg^{-1}$, this corresponds to $7.35\,GJ\,ha^{-1}$ or $73.5\,J\,cm^{-2}$.

Species with protected meristems, such as Shepherd's purse (*Capsella bursa-pastoris* L.), were tolerant due to regrowth after flaming, and they could only be completely killed in their early stages. Annual bluegrass (*Poa annua* L.) could not be completely killed with a single flame treatment, regardless of developmental stage or propane dose. Considerably, lower doses (40%) were required in years with higher precipitation compared with a dry year. Because precipitation enhances thermal weed control efficacy, a system that induces high humidity could provide better weed control.

3.2.2 Steam Treatment

Steam-based weed control has received renewed interest in recent years. The most common and simplest steam applicator is sheet steaming. This involves covering the soil with

TABLE 3.1   Parameter Estimates of Regression Model for Plant Number Data After Flame Treatment of White Mustard at Different Plant Sizes and Densities

Number of Leaves	Plant Density (No. m^{-2})	D From Eq. (3.1) (No. m^{-2})	$a = LD_{50}$ (kg ha^{-1})	b From Eq. (3.1)	LD$_{95}$ (kg ha^{-1})	LD$_{95}$ (GJ ha^{-1})
0–2	195	174	21.7	4.55	41.5	2.03
0–2	395	342	21.7	4.55	41.5	2.03
2–4	169	155	38.8	4.55	74.1	3.63
2–4	365	335	38.8	4.55	74.1	3.63
0–2	250	207	22.1	3.87	47.3	2.31
0–2	714	658	22.1	3.87	47.3	2.31
3–4	265	210	35.6	4.76	66.1	3.24
3–4	798	607	60.4	3.01	159.5	7.82

Modified from Ascard, J., 1994. Dose-response models for flame weeding in relation to plant size and density. Weed Res. 34 (5), 377–385.

a thermally resistant membrane, which is sealed at the edges. Steam, which is pumped under the sheet, penetrates the soil's surface layer to kill weeds and their seeds (Gay et al., 2010b). A more mobile option is to use a small hooded applicator head, connected to a steam source via a hose, and to apply saturated steam to the soil surface (Gay et al., 2010b).

Gay et al. (2010b) tested a hooded applicator with an area of 150×150 mm. Their steam generator had a nominal duty cycle of 8.5 kW with a 1.6 kW superheater to deliver 4 kg h^{-1} of superheated steam. It heated an area of soil (100×50 mm) to a temperature of 90°C or more in 200 s. After 300 s, the entire soil surface under the applicator reached a uniform temperature of 100°C (Gay et al., 2010b). Given that the thermal capacity of water is 4.2 kJ kg^{-1}°C^{-1} and the latent heat of vaporization for water is 2.26 MJ kg^{-1} and assuming an initial water temperature of 25°C, this represents an application energy of 13.5 kJ cm^{-2}.

Raffaelli et al. (2016) developed a band steaming system for field work. In their investigation of the system's performance, they used the following relationship for weed survival as a function of steam application:

$$Y = \frac{D - C}{1 + e^{\{b[\log(x) - \log(LD_{50})]\}}} + C \tag{3.2}$$

where Y is the weed response (plants m^{-2}), D is the upper limit of response (plants m^{-2}), C is the lower limit of response (plants m^{-2}), x is the applied steam (kg m^{-2}), and LD$_{50}$ is the applied steam needed to achieve a 50% weed mortality rate (kg m^{-2}).

The parameter values from their experimental trials with the system were $b = 2.6$, $C = -2.8$, $D = 72.5$, and LD$_{50} = 1.0$ (kg m^{-2}) (Raffaelli et al., 2016). Based on these values, the LD$_{90}$ (90% weed control) for this system was 2.3 kg m^{-2} of steam. Based on the earlier assumptions about steam production, achieving the LD$_{90}$ dose rate required 592 kJ cm^{-2}.

Soil pasteurization can also be achieved by injecting steam into the soil using gridded steam injectors (Gay et al., 2010a,b). This technique could be used as an alternative to soil fumigation, which is commonly applied in high-value horticultural crops.

Gay et al. (2010b) developed a scalar index to measure heating efficiency for steam soil heating:

$$I = \frac{1}{Vt_f} \int_0^{t_f} \int_V T \cdot dV \cdot dt \tag{3.3}$$

where V is the volume of soil being heated (m^3), T is the temperature increase (K), and t_f is the heating time (s). Gay et al. (2010b) applied this index to their steam experiments. Eq. (3.3) represents the 4-D average of the temperature change in the soil volume. The performance of sheet steaming varied between 7.5 and 18, the hooded applicator varied between 27 and 37, and the steam injection system varied between 37 and 47 (Gay et al., 2010b).

While steam treatment is effective at killing weeds and can achieve some pasteurization of the soil (Gay et al., 2010a,b), it requires considerable energy investment to create the steam. This is partly due to the inherent limitations of convective heat transfer.

3.2.3 Convective Heating

Thermal weed control depends on heat transfer from a hot fluid (air, water, or steam) into the weed plants. The heat flow from a fluid with a temperature of T_f to a solid with a temperature of T_s is expressed as

$$\frac{q}{A} = h(T_s - T_f) \tag{3.4}$$

where h is the convective heat flow coefficient of the material's surface (Holman, 1997). When studying thermodynamic processes, temperatures are usually expressed in absolute (Kelvin) values.

This convective heat flow coefficient depends on a number of other parameters and conditions (Welty et al., 2007). For example, the convective heat flow coefficient for a vertical surface where natural convection achieves turbulent flow conditions over the surface is given by (Welty et al., 2007) the following:

$$h = \frac{k}{L} \left\{ 0.825 + \frac{0.387 Ra_L^{1/6}}{\left[1 + \left(\frac{0.492}{\mathrm{Pr}} \right)^{9/16} \right]^{8/27}} \right\} \tag{3.5}$$

where k is the thermal conductivity of the heating fluid (W m^{-1}K^{-1}) and L is the characteristic length of the object being heated (m).

The Rayleigh number (Ra_L) in this equation is also based on a complex relationship between temperature and the physical properties of the fluid. It is given by (Welty et al., 2007) the following:

$$Ra_L = \frac{g\beta}{\nu\alpha}(T_s - T_\infty)L^3 \tag{3.6}$$

where g is the acceleration due to gravity, β is the thermal expansion coefficient of the fluid, ν is the kinematic viscosity of the fluid medium, α is the thermal diffusivity of the fluid medium, and L is the characteristic length of the surface.

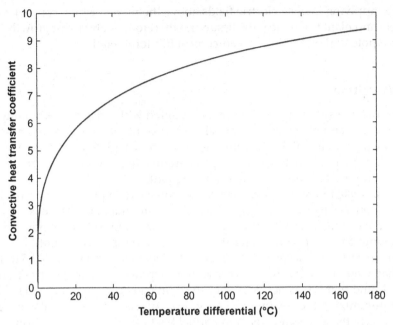

FIG. 3.1 Convective heat transfer coefficient (h) for air as a function of temperature differential between an object and the air.

Finally, the Prandtl number used in Eq. (3.5) is a relationship between the fluid's viscous and thermal diffusion rates given by (Welty et al., 2007) the following:

$$Pr = \frac{\nu}{\alpha} \tag{3.7}$$

Close examination of these equations shows that the convective heat transfer coefficient is dependent on the temperature differential between the fluid and the surface of the material (see Fig. 3.1). An important consequence of this temperature dependence is that heat transfer diminishes rapidly as the temperature differential between the object and the surrounding fluid reduces; therefore, the efficacy of thermal weeding is limited by the rate of heat transfer into the plants. It could be better if the weed control system could generate heat directly in the plant or soil itself, rather than relying on heat transfer.

It is also evident that there is an upper limit for convective heat transfer, which becomes independent of very high-temperature differences (Fig. 3.1); therefore, a system that directly generates heat in the weeds could be more effective than steam treatment.

3.3 ELECTRICAL SYSTEMS

Electric weed control falls into two main categories: one form involves applying high voltage directly to plants to generate current flow through the plant in accordance with Ohm's law (Smith, 1976), which generates heat in the current path through the plants; the other involves applying intense electric fields above the plants, which concentrates the electric field at leaf and plant tips to heat the tissue and therefore inhibit plant growth. This section explores these potential methods of weed control in more detail.

3.3.1 Electrocution

Weed electrocution systems have been experimented with since the 19th century. They involve the use of a high-voltage alternating electric current, passing through a plant, to generate heat in the plant material due to ohmic heating. These high temperatures vaporize water contained within the plant, increasing the pressure inside plant cells, which causes the cell membranes to rupture and consequently kill the plant.

Diprose et al. (1980) found that high alternating voltages (frequency unspecified) at around 5 kV (RMS) can physically destroy annual beet in pots in under 20 s. The electric current was applied by two electrodes; one is a brass rod pushed into the soil in the pot and the other a strip of aluminum foil wrapped around the foliage at the top of the plant.

Maximum currents vary according to the plant species and range from 0.5 to 1.0 A (RMS) for plants that were 1.0–1.4 m high, representing a power requirement between 2.5 and 5.0 kW. They found that it was not necessary to completely burn the plant to kill it, and various ways of minimizing the treatment energy were suggested. Field trials, using handheld electrodes to apply the electric currents to annual beet growing among a crop, showed that much larger power (up to 20 kW) was necessary to kill the plants. A tractor-driven system was constructed to produce an alternating voltage of 8 kV (RMS). This system enabled it to cover six rows and travel at speeds up to 16 km h^{-1}. It effectively killed 75% of the bolting and weed beet.

Compared with chemical or mechanical weed control, this method does not introduce any chemicals into the food chain and does not disturb the soil surface; however, electrocution requires very high voltages and is very time-consuming. This introduces an operational health and safety hazard and has not been widely adopted.

3.3.2 Electrostatic Fields

Somewhat like electrocution, which is a direct contact approach, research into the effect of noncontact high-intensity electrostatic fields on plant growth was conducted from the mid-18th, 20th century. Early experiments indicated that increased yields from both cereal and vegetable crops could be obtained by applying electrostatic fields to plants while they were growing (Newman, 1911; Jorgensen and Stiles, 1917; Hendrick, 1918; Shibusawa and Shibata, 1930); however, this beneficial effect is no longer accepted.

Murr (1963a,b) undertook experiments where an electrostatic field was established between two aluminum wire grids. A lower electrode was placed in the soil and orchard grass (*Dactylis glomerata*) was used as a test species. An upper electrode was suspended above the soil and adjusted in height to vary the electric field strength, although the upper grid was never >10 cm above the tops of the plants.

Temperature and light intensity were controlled (unspecified), and 16 h day length was used. The control plots had the same electrode arrangement as the "active" ones, but without any applied voltage. The top electrode maintained a positive voltage relative to the bottom one, simulating the earth's natural fair weather electric field (Kelley, 2013). Murr (1963a,b) observed that during continuous exposure to the electrostatic fields, the leaf tips of the seedlings began to brown, as if burnt, and he noted the similarity to mineral deficiency symptoms.

Murr defined a damage factor as the proportion of the dry weight of electrified material compared with control samples. For orchard grass, damage was found to rise to 25% when the electrostatic field strength was $50\,kV\,m^{-1}$ and then rapidly increased to 50% at $75\,kV\,m^{-1}$. Similar results were obtained with seedlings of reed canary grass (*Phalaris arundinacea*) (Murr, 1963b). Transverse sections of some leaves showed that the epidermal cells had been destroyed in the browned tip and damaged in the dark green zones, which bordered the browned tips. There was complete absence of cell structure in the tip area and chloroplast derangement in the dark green band.

After observing the physical characteristics of the damage, Bachman and Reichmanis (1973) explored the growth rates of plants that were subjected to electrostatic fields. By using both sloping and horizontal upper electrodes, they found that barley seedlings grew until they were about 2 cm from the electrode, so that the plant tops followed the profile of the electrode. At this point, the field strength corresponded to about $800\,kV\,m^{-1}$. Similarly, growth rates above $200\,kV\,m^{-1}$ were found to be inhibited when compared with controls, with 300 and $400\,kV\,m^{-1}$ having stronger effects (Bachman and Reichmanis, 1973). By extrapolating these growth rate results, they predicted that a zero value should occur at approximately $800\,kV\,m^{-1}$, which agreed with their earlier findings.

Although plants have been killed by the application of very high electrostatic fields, it is difficult to use this method for weed control. This method requires arrays of wires suspended some 2 m or more above crops, and all voltages are in the tens of kV. Outside of a laboratory, this system is cumbersome and dangerous due to its very high field strength. It would be better to use a system that required lower field strength.

3.4 ELECTROMAGNETIC FIELDS

While electrostatic fields have some effect on plants, a time-changing field can transfer energy into plants using far less intense fields. A time-changing magnetic field will induce a changing electric field and vice versa. These changing fields form electromagnetic waves. Electromagnetic waves are a complex phenomenon because they can propagate through vacuum without the need for a material medium, they simultaneously behave like waves and like particles (Dirac, 1927; Einstein, 1951), and they are intrinsically linked to the behavior of the space-time continuum (Einstein, 1916). It can be shown that magnetic fields appear because of relativistic motion of electric fields, which is why electricity and magnetism are so closely linked (Chappell et al., 2010). It has even been suggested that electromagnetic phenomena may be a space-time phenomenon, with gravitation being the result of space-time curvature (Einstein, 1916) and electromagnetic behavior being the result of space-time torsion (Evans, 2005).

An electromagnetic wave is described in terms of its

1. frequency (*f*), which is the number of waves that pass a fixed point in an interval of time. Frequencies are usually measured as waves per second or cycles per second, which is given the unit of hertz (Hz);

2. wavelength (λ), which is the distance between successive crests or troughs in the wave. If frequencies are measured in hertz, then wavelengths are measured in meters (m);
3. speed (c), which is measured in meters per second and is determined by the electric and magnetic properties of the space through which the wave travels.

Electromagnetic waves can be of any frequency; therefore, the full range of possible frequencies is referred to as the electromagnetic spectrum. Although Maxwell's equations, which describe all electromagnetic phenomena, do not indicate any limits on the spectrum, the known electromagnetic spectrum extends from frequencies of around $f = 3 \times 10^3$ Hz ($\lambda = 100$ km) to $f = 3 \times 10^{26}$ Hz ($\lambda = 10^{-18}$ m). This covers everything from ultralong radio waves to high-energy gamma rays (International Telecommunication Union, 2004).

Electromagnetic waves can transfer energy from one object to another through open space. Generally, the amount of energy transferred depends on the intensity of the electromagnetic fields, the frequency of the fields' oscillations, and the dielectric properties of the material. Therefore, quite moderate fields can transfer high energy, if the frequency of the oscillation and the dielectric properties are both very high.

A dielectric material, such as a weed plant, exhibits both polarization and direct conduction within the material as charges are displaced by the applied fields (Metaxas and Meredith, 1983). The combination of polarization and conduction gives rise to displacement currents within the material. These currents are not in phase with the applied electromagnetic field. This phase difference is described in mathematical terms by using complex numbers. The complex displacement current can be resolved into a reactive component (based on the imaginary part of the complex number) and a real component. To determine the current density in a dielectric material using Maxwell's equations and the physics of dielectric materials requires the introduction of a complex dielectric permittivity $\varepsilon^* = \varepsilon' + j\,\varepsilon''$ (Debye, 1929).

The real part of the dielectric permittivity ε' expresses the material's ability to store electric energy (Singh and Heldman, 1993) and thus represents the reactive nature of the material's dielectric properties (Smith, 1976). In particular, ε' influences the wave impudence of the space occupied by the dielectric causing a reduction of the propagation speed of light in the material, refraction of the wave through the material, and reflections at the interfacial boundary between the air and the dielectric material (Giancoli, 1989; Montoro et al., 1999). The change in propagating speed reduces the wavelength of the electromagnetic waves inside the dielectric material.

The imaginary part of the dielectric permittivity ε'', is known as the loss factor and represents the resistive nature of the material's dielectric properties (Smith, 1976; Giancoli, 1989). Resistive losses within the medium reduce the amplitude of the electromagnetic waves and generate heat inside the material.

It is common practice to express the dielectric properties of a material in terms of the relative dielectric constants κ' and κ'', which are defined such that $\varepsilon' = \kappa' \varepsilon_o$ and $\varepsilon'' = \kappa'' \varepsilon_o$, where ε_o is the permittivity for free space. In summary, when an electromagnetic wave is transmitted through a dielectric material, there will be attenuation and delay in the signal compared with a wave traveling through free space. The dielectric properties of vegetation are significantly higher at microwave frequencies (from 300 MHz to 300 GHz) than at higher frequencies, including the infrared, visible, and ultra violet parts of the spectrum (Fig. 3.2); therefore, microwave energy offers a potentially better option for weed treatment than other electromagnetic waves. Moisture content of the vegetation profoundly affects these properties as well (Fig. 3.2).

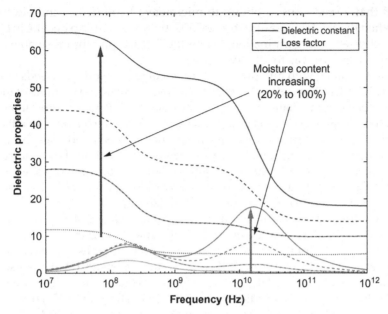

FIG. 3.2   Dielectric properties of vegetation as a function of frequency and moisture content (based on dielectric models for vegetation, bound water, and free water, developed by Ulaby and El-Rayes, 1987; Meissner and Wentz, 2004; Serdyuk, 2008).

3.4.1 Microwave Weed Control

Interest in the effects of high-frequency electromagnetic waves on biological materials dates back to the late 19th century (Ark and Parry, 1940), while interest in the effect of high-frequency waves on plant material began in the 1920s (Ark and Parry, 1940). Many of the earlier experiments on plant material focused on the effect of radio frequencies (RF) on seeds (Ark and Parry, 1940). In many cases, short exposure resulted in increased germination and vigor of the emerging seedlings (Tran, 1979; Nelson and Stetson, 1985); however, long exposure usually resulted in seed death (Ark and Parry, 1940; Bebawi et al., 2007; Brodie et al., 2009).

Davis et al. (1971, 1973) were among the first to study the lethal effect of microwave heating on seeds. They treated seeds, with and without any soil, in a microwave oven and showed that seed damage was mostly influenced by a combination of seed moisture content and the energy absorbed per seed. Other findings from the study by Davis et al. (1971) suggested that both the specific mass and specific volume of the seeds were strongly related to a seed's susceptibility to damage by microwave fields (Davis et al., 1973). The association between the seed's volume and its susceptibility to microwave treatment may be linked to the *radar cross section* (Wolf et al., 1993) presented by seeds to propagating microwaves. Large radar cross sections allow the seeds to intercept and, therefore, absorb more microwave energy.

Barker and Craker (1991) investigated the use of microwave heating in soils of varying moisture content (10–280 g water/kg of soil) to kill "Ogle" oat (*Avena sativa*) seeds and an undefined number of naturalized weed seeds present in their soil samples. Their results demonstrated that a seed's susceptibility to microwave treatment is entirely temperature-dependent. When the soil temperature rose to 75°C, there was a sharp decline in both oat seed and naturalized weed seed germination. When the soil temperature rose above 80°C, seed germination in all species was totally inhibited.

Several patents dealing with microwave treatment of weeds and their seeds have been registered (Haller, 2002; Clark and Kissell, 2003; Grigorov, 2003); however, none of these systems appear to have been commercially developed. This may be due to concerns about the energy requirements to manage weed seeds in the soil using microwave energy. In a theoretical argument based on the dielectric and density properties of seeds and soils, Nelson (Nelson, 1996) demonstrated that using microwaves to selectively heat seeds in the soil *cannot be expected*. He concluded that seed susceptibility to damage from microwave treatment is a purely thermal effect, resulting from soil heating and thermal conduction into the seeds. This has been confirmed experimentally by Brodie et al. (2007a).

Experience confirms that microwave energy can kill a range of weed seeds in the soil (Davis et al., 1971, 1973; Barker and Craker, 1991; Brodie et al., 2009).

Presowing microwave irradiation of soil minimizes weed establishment (Davis et al., 1971, 1973; Sartorato et al., 2006; Brodie et al., 2012; Brodie and Hollins, 2015). It can also destroy the weed reproductive plant parts and their seeds that are covered up by soil at a depth of several centimeters (Diprose et al., 1984; Brodie et al., 2007b). Wayland et al. (1973) treated wheat and radish seeds in situ at 25 mm depth, and moisture content of soil was 6.5%. They found that microwave treatment was toxic to seeds with a threshold of 10 J cm^{-2} of energy density. Increasing power density was more effective at reducing the germination percentage of seeds than simply increasing energy density (exposure time at a fixed power level) for some species.

Davis et al. (1971) conducted an experiment to evaluate the effect of microwave treatment on the seedling survival percentage of 12 species. They described that the 48 h germinated seedling showed no survival after a short exposure of microwave energy and concluded that susceptibility of young seedlings to microwave heating was highly correlated with moisture content and absorption of energy. Menges and Wayland (1974) compared postemergence herbicides (metazoal, propachlor, and perfluidone) with microwave at energy density of 45–720 J cm^{-2} for weed suppression in an onion crop. They reported that microwave (360 J cm^{-2}) irradiation significantly inhibited weeds establishment. Additionally, minimum crop injury was noted in the case of microwave treatment (18%) compared with herbicide application (85%).

Brodie (Brodie et al., 2007a,b,c, 2009; Brodie and Hollins, 2015) has conducted several weed seed experiments, where either air dry or moist soil (20% moisture by volume) was layered into pots with sets of between 10 and 25 seeds placed into paper envelopes at depths of 0, 2, 5, 10, and 20 cm within each pot. Pots were treated for 0, 2, 5, 10, 30, 60, or 120 s using a prototype microwave system fed from a conventional 600 W microwave oven magnetron into a pyramidal horn antenna (Fig. 3.3). The paper envelopes allowed easy seed recovery from the soil after treatment so that seeds could be germinated in a growth cabinet for viability assessment. The following species were evaluated: annual ryegrass (*Lolium rigidum* L.); perennial ryegrass; bellyache bush (*Jatropha gossypiifolia* L.); giant sensitive tree, catclaw plant, or bashful plant (*Mimosa pigra* L.); *Parthenium* (*Parthenium hysterophorus* L.); rubber vine (*Cryptostegia grandiflora* R.Br.); wild radish (*Raphanus raphanistrum* L.); and wild oats (*Avena fatua* L.).

FIG. 3.3 Microwave prototype system based on a modified microwave oven, including wooden box used for soil bacteria study. *Data from Bebawi, F.F., Cooper, A.P., Brodie, G.I., Madigan, B.A., Vitelli, J.S., Worsley, K.J., Davis, K. M., 2007. Effect of microwave radiation on seed mortality of rubber vine* (Cryptostegia grandiflora r. Br.), *parthenium* (Parthenium hysterophorous L.) *and bellyache bush* (Jatropha gossypiifolia L.). *Plant Prot. Q. 22 (4), 136–142.*

Plants were treated with the same microwave apparatus (Fig. 3.3) for 0, 2, 5, 10, 30, or 120 s and evaluated 5 days after treatment to determine the number of living and dead plants. Experiments have been conducted for the following species: annual ryegrass (*Lolium multiflorum* L.), barley grass (*Hordeum vulgare* L.), barnyard grass (*Echinochloa crus-galli*), fleabane (*Conyza bonariensis* L.), marshmallow (*Malva parviflora* L.), prickly paddy melon (*Cucumis myriocarpus*), wild radish (*Raphanus raphanistrum*), and wild oat (*Avena fatua*) (Brodie et al., 2007b, 2012; Brodie and Hollins, 2015). In each case, both pot trials and in situ field experiments were conducted for each species.

Assuming that plant responses are normally distributed, the data were fitted to a Gaussian error function response curve, which is the integral of the normal distribution response:

$$S = a \cdot \text{erfc}[b(\Psi - c)] \tag{3.8}$$

where S is the normalized survival rate for plants; Ψ is the estimated microwave energy at ground level ($J\,cm^{-2}$); and a, b, and c are constants to be experimentally determined for each species.

TABLE 3.2 Equation Coefficients, Goodness of Fit (R^2), LD_{50}, and LD_{90} for Weed Plant Survival as a Function of Microwave Energy Applied to the Soil Surface

Species	Coefficients						R^2	LD_{50} ($J\,cm^{-2}$)	LD_{90} ($J\,cm^{-2}$)
	a	b	c	d	e	f			
Annual ryegrass	−12.7	0.01757	12.65	15.78	0.016	3.296	0.72	15.5	110
Barnyard grass	0.66	0.03268	16.58	–	–		0.98	23	48
Fleabane	0.528	0.1313	10.45	–	–		1.00	11	18
Marshmallow	0.553	0.02885	33.55	–	–		0.98	36	66
Paddy melon	0.553	0.1033	10.26	–	–		0.99	11	19
Snails	1.034	0.0824	0	–	–		0.74	6	14
Wild radish	0.523	0.04028	27.83	–	–		1.00	29	51

Note: Rows with blank cells indicate data models based on Eq. (3.8).

Table 3.2 shows the dose-response parameters for all emerged plant species. The dose responses of these data were modeled by Eq. (3.8); however, in the case of annual ryegrass, the relationship between applied microwave energy and plant survival was better described by

$$S = a \cdot \text{erfc}[b(\varPsi - c)] + d \cdot \text{erfc}[e(\varPsi - f)] \tag{3.9}$$

The relationships between applied microwave energy and seed survival were fitted to a dose-response surface of the form:

$$S = a \cdot \text{erfc}\left[b \cdot \left(\varPsi \cdot e^{-2cd} - f\right)\right] \tag{3.10}$$

where d is the depth of the seeds in the soil profile (m) and a, b, c (field attenuation rate in soil), and f (median seed response) are constants to be experimentally determined for each species. The various equation coefficients and goodness of fit (R^2) for weed seed responses are listed in Table 3.3.

Speir et al. (1986) examined the effect of microwave energy on low fertility soil (100 randomly selected cores at depth of 50 mm), microbial biomass, nitrogen, phosphorus, and phosphatase activity. They reported that an increase in microwave treatment duration (90 s) dramatically increased the nitrogen level in the soil (106 µg N g^{-1}), but available phosphorus concentration declined as treatment time increased. This is also consistent with the findings of Gibson et al. (1988), who demonstrated that shoot and root growth of birch (*Betula pendula*) was significantly increased in microwave irradiated soil. Their experiment was to evaluate the effect of microwave treatment of soil supplemented with two mycorrhizae on birch seedlings. Shoot growth progressively increased with irradiation duration, with the highest dry shoot weight being 84 mg for their highest irradiation duration (120 s) compared with nonirradiated soil (25 mg). This result was with no mycorrhizal supplementation.

A fully replicated pot experiment (five pots per treatment with three plants per pot) was undertaken to explore the effect of graduated microwave treatments (applied energy at the soil surface = 0, 60, 140, and 240 J cm^{-2}) on plant growth and final yield per pot of wheat (*Triticum* spp.), rice (*Oryza sativa*), maize (*Zea mays*), and canola (*Brassica napus*). The soil was irradiated while at field capacity, in terms of its moisture status. The crop seeds were

TABLE 3.3 Equation Coefficients and Goodness of Fit (R^2) for Weed Seed Survival as a Function of Microwave Energy Applied to the Soil Surface, Seed Burial Depth, and Soil Moisture Status

Species	Dry					Wet				
	a	b	c	f	R^2	a	b	c	f	R^2
Annual ryegrass	0.31	0.0145	0.06	340.2	0.52	0.30	0.1522	0.07	81.42	0.91
Bellyache bush	0.50	0.0100	0.08	133.8	0.90	1.03	0.0085	0.23	−6.266	0.83
Parthenium	0.55	0.0051	0.22	170.8	0.79	0.96	0.0055	0.04	0.0014	0.73
Perennial ryegrass	0.41	0.0121	0.06	313.1	0.78	0.43	0.0660	0.13	53.78	0.94
Rubber vine	0.49	0.0090	0.03	209.4	0.58	0.61	0.0069	0.03	90.75	0.86
Wild oats	0.46	0.0124	0.12	225	0.76	0.82	0.0056	0.09	1.493	0.74
Wild radish	–	–	–	–	–	0.16	0.2415	0.12	32.72	0.72

TABLE 3.4 Mean Crop Response as a Function of Applied Microwave Treatment Energy

| Species | Microwave Treatment (J cm^{-2}) | | | | | LSD$_{5\%}$ | Change From Hand-Weeded Control |
	0	0 (Hand Weeded)	60	140	240		
Canola pod yield (g pot^{-1})	0.27^a	0.56^a	0.36^a	1.25^b	1.95^c	0.55	250%
Days to flowering—canola	71.4^a	67.6^{ab}	70.2^a	63.2^b	61^b	7.10	14.6%
Maize (g pot^{-1})	5.25^a	6.63^a	–	10.28^{ab}	12.76^b	4.80	92%
Rice grain yield (g pot^{-1})	40.00^a	41.3^a	43.25^a	59.00^{ab}	64.00^b	18.90	55%
Wheat grain yield (g pot^{-1})	0.66^a	0.67^a	0.68^a	0.75^a	1.25^b	0.30	87%

Note: Entries with different superscripts across the rows are statistically different from one another.

planted 1 day after microwave treatment, to ensure that the soil had returned to ambient equilibrium conditions. Plant growth rate (canola flowing dates in Table 3.4), final plant height, and crop yield were increased with the increasing level of applied microwave energy (Table 3.4). A hand-weeded control was also included in these experiments, so that microwave soil treatment could be compared with an optimal weed control scenario. Microwave soil treatment has significant beneficial effects on subsequent crop growth, beyond the simple control of weed competition.

Using equations derived for the temperature distribution in microwave heated soil (Brodie et al., 2007b; Brodie, 2008), the performance index, which was developed by Gay et al. (2010b), can be derived. After substituting the temperature distribution equations, developed by Brodie, into Eq. (3.3), the performance of microwave soil treatment using a horn antenna to irradiate the soil is given by

$$I = \frac{n\omega\varepsilon_o\kappa'' E_o^2}{128 t_f k \gamma \alpha^4} \left(e^{4\gamma\alpha^2 t_f} - 1 \right) \tag{3.11}$$

where n is an amplitude scaling factor for simultaneous heat and moisture movement during microwave heating (Henry, 1948; Brodie, 2007), ω is the angular velocity of the microwave field (rad s^{-1}), ε_o is the permittivity of free space, κ'' is the dielectric loss factor of the soil, E_o is the electric field strength at the soil surface (V m^{-1}), t_f is the microwave heating time (s), k is the thermal conductivity of the soil (W m^{-1}K^{-1}), γ is the combined heat and moisture diffusivity coefficient for the soil, and α is the microwave attenuation factor in the soil (m^{-1}). Eq. (3.11) can be used to estimate the heating efficiency index for clay soil, which is being heated for 120 s using a 200 W, 2.45 GHz microwave source feeding into a horn antenna with aperture dimensions 110 × 55 mm. The heating efficiency index is 80, which is almost twice that of steam treatment, as estimated by Gay et al. (2010b). This improvement in the heating efficiency index is probably linked to the rapidity of microwave heating, compared with steam treatment, because microwave heating is a spatial energy transfer phenomenon while steam heating depends on convective heat transfer.

3.4.2 Infrared Radiation

Heat kills plants, there being a time-temperature relationship (Levitt, 1980). Radiative heat transfer refers to the transfer of energy by broad-spectrum electromagnetic radiation from some adjacent hot object (or from a hot environment) to the heated object. Any object that is above 0 K will radiate energy in the form of electromagnetic photons. The German physicist, Max Planck (1858–1947), deduced that the radiation spectral density (ρ) given off from a hot object depended on the wavelength of interest and the temperature of the object. This spectral density can be described by

$$\rho = \frac{2hc^2}{\lambda^5 \left\{ e^{\frac{hc}{\lambda k T}} - 1 \right\}} \tag{3.12}$$

where h is Planck's constant (6.6256×10^{-34} J s), c is the speed of light, λ is the electromagnetic wavelength of interest, k is Boltzmann's constant (1.38054×10^{-23} J K^{-1}), and T is the temperature in Kelvin. A typical set of spectral distributions for different temperatures is shown in Fig. 3.4.

The brightness temperature of a body can be determined by rearranging Planck's equation to find T for a given spectral density value:

$$T = \frac{hc^2}{\lambda k \cdot \ln\left(\frac{2\pi hc}{\rho \lambda^5} + 1\right)} \tag{3.13}$$

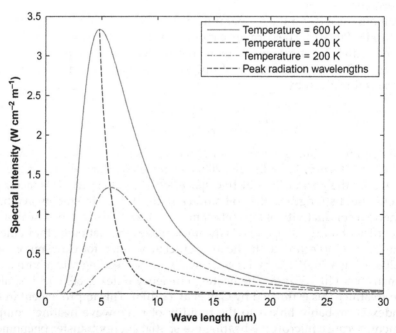

FIG. 3.4 Radiative spectral density at different temperatures as a function of temperature and wavelength.

The wavelength at which peak radiation intensity occurs can be found by differentiating Planck's equation and setting the derivative equal to zero (Appendix A). Therefore, the wavelength of peak radiation is determined by

$$\lambda_p \approx \frac{hc}{5kT} \tag{3.14}$$

where λ_p is the peak radiation wavelength (m). At room temperature or above, the wavelength of peak radiation will be in the micrometer range ($\sim 10\,\mu m$), which is in the long-wavelength infrared band (Table 3.5). The penetration of electromagnetic energy into materials is limited by the wavelength and the dielectric properties of the material (Vollmer, 2004):

$$\delta = \frac{\lambda_p}{4\pi\sqrt{\kappa}} \tag{3.15}$$

where δ is the penetration depth (m) and κ is the relative dielectric constant of the material. The penetration depth of any radiation from objects at room temperature or above will be in the nanometer range; therefore, radiative heat transfer must be regarded as a surface phenomenon where further heat transfer from the surface into the material occurs via internal conduction and convection.

The total radiated power can be determined by integrating Planck's equation across all wavelengths for a particular temperature (Appendix B) to yield the Stefan-Boltzmann equation. The power transferred from an object at one temperature to another object at a lower temperature is given by (Holman, 1997):

$$q = \varepsilon\sigma A\left(T_A^4 - T_p^4\right) \tag{3.16}$$

where q is the radiation power (W), ε is the surface emissivity of the radiator material, σ is the Stefan-Boltzmann constant ($5.6704 \times 10^{-8}\,\mathrm{J\,s^{-1}\,m^{-2}\,K^{-4}}$), A is the surface area of the heated object (m^2), T_A is the temperature of the infrared applicator (K), and T_p is the temperature of the plants being treated (K).

Denaturing of plant cell components starts with long-term exposure to temperatures of about 40°C. The fatal impacts of high temperatures on plants have been studied in detail for over a century (Levitt, 1980). In particular, a thoroughly demonstrated empirical relationship between lethal temperature and temperature holding time has been developed by Lepeschkin (1912):

$$T = 79.8 - 12.8 \cdot \log_{10} Z \tag{3.17}$$

TABLE 3.5 A Commonly Used Subdivision Scheme

Division Name	Abbreviation	Wavelength (μm)	Temperature (K)
Near infrared	NIR	0.75–1.4	3964–2070
Short-wavelength Infrared	SWIR	1.4–3.0	2070–966
Mid-wavelength infrared	MWIR	3.0–8.0	966–362
Long-wavelength infrared	LWIR	8.0–15.0	362–193
Far infrared	FIR	15.0–1000	193–3

TABLE 3.6 Lethal Temperatures as a Function of Exposure Time

Required Heat Holding Time (s)	Required Lethal Temperature (°C)
2.0	98
1.0	102
0.1	115
0.05	119
0.01	128

Data from Levitt, J., 1980. Response of Plants to Environmental Stresses, vol. 1. Academic Press, New York.

where T is the lethal temperature (°C) and Z is the lethal temperature holding time, in minutes (Levitt, 1980).

Although there will be some variability in lethal temperature between individuals within a species and between species, literature suggests that this basic relationship holds in almost all cases. From this simple relationship, the lethal temperature for plants depends on holding time. Table 3.6 illustrates the lethal temperatures required for various treatment times.

Infrared radiation systems use gas burners to heat ceramic and metal surfaces, which then radiate infrared energy toward the ground. According to Parish (1990), laboratory investigations identified that a "medium-wave tubular fused quartz infrared emitter" was the most effective for weed control. Infrared burners are not affected by wind, in contrast to flame weeders, and they cover a more closely defined area.

Ascard (1998) discovered that efficacy of flaming and infrared radiation treatment, on emerging seedlings, was similar. For example, when white mustard (*Sinapis alba* L.) plants were at the four-leaf stage, propane doses of $8\,kg\,ha^{-1}$ from either flaming or infrared systems merely scorched the edges of the leaves. Propane doses of $30\,kg\,ha^{-1}$ desiccated almost 20% of the plants, but surviving plants showed vigorous regrowth. One hundred percent weed control required $120\,kg\,ha^{-1}$ of propane for both systems.

Considerably higher temperatures were required under the flamer compared with the infrared radiator. Temperatures of up to 1350°C were recorded in the central blue part of the stationary flamer system; however, the stationary infrared radiator had a maximum temperature of 770°C (Ascard, 1998). The ground temperature in both cases was approximately 180° C (Ascard, 1998). Ascard (1998) also reports work by Hoffmann who found that infrared radiators cause a higher temperature increase in the upper few millimeters of soil compared with flamers, because radiation heating avoids the convective heat transfer limitations, which are associated with hot air (flame) heating.

For efficient plant destruction, an infrared radiator is required, which produces high energy intensity at a wavelength that is absorbed, rather than reflected or transmitted, by the plant tissues. To kill young white mustard plants, an energy density at ground level of between 200 and $400\,kJ\,m^{-2}$ (or 20–$40\,J\,cm^{-2}$) of short-wave or medium-wave infrared energy is required to severely restrict plant growth (Parish, 1990). These dose rates are similar to those associated with microwave weed control discussed in the previous section; however, because microwaves have a much longer wavelength than infrared radiation, the penetration of microwave energy into plants and the soil will be much further.

In several studies, infrared radiators have proved to be inferior compared with flame weeders, but Ascard (1998) and Parish (1990) found that the differences in effect were dependent on the type of thermal weeder, dose, ground speed, burner height, plant size, plant density, and plant species. Their studies also indicate that infrared burners are more likely to suffer from shading interference in dense vegetation compared with flame weeders that cause turbulence and thereby expose more leaves to the flame. This shading effect is linked to the shallow penetration of infrared radiation into most dielectric materials. Because microwave energy has a much longer wavelength than infrared energy, microwave weed control is less vulnerable to shading, than other radiation systems, including ultraviolet radiation.

3.4.3 Ultraviolet Radiation

The wavelength of ultraviolet (UV) radiation lies between 100 and 400 nm and is thus outside the visible range. UV rays can be separated into three groups on the basis of wavelength: UV-A (320–400 nm), UV-B (280–320 nm), and UV-C (100–280 nm). When plants are irradiated with UV, almost all energy is absorbed in the outermost 0.1–0.2 mm layer of the plant tissue. This results in heating of the plant tissue and thus can have effects similar to the damage to plants from flame weeding (Andreasen et al., 1999).

Andreasen et al. (1999) irradiated four weed species at two different leaf stages and two crops at one-leaf stage with ultraviolet light from a water cooled 2.35 kW UV lamp. The weed species were annual bluegrass (*Poa annua* L.), common groundsel (*Senecio vulgaris* L.), shepherd's purse (*Capsella bursa-pastoris* (L.) Medicus), and small nettle (*Urtica urens* L.). The crop species were canola (*Brassica napus* L. ssp. napus) and pea (*Pisum sativum* L.). Plants were treated in a laboratory with the UV lamp placed as close as possible to the plant canopy without touching it (about 1 cm above). Plant parts close to the lamp received more radiation than parts farther away. After irradiation, the aboveground fresh weight was measured after the plants were withered but before regrowth commenced from undamaged buds.

Andreasen et al. (1999) used the following model to fit their data:

$$Y = \frac{D-C}{1+e^{\{b[\log(x)-\log(\mathrm{LD}_{50})]\}}} + C \qquad (3.18)$$

where Y is the fresh weight yield to a UV dose of x (GJ ha^{-1}). D is the upper limit of fresh weight, C is the lower limit (g pot^{-1}), and b and LD_{50} are determined experimentally. Table 3.7 lists the dose-response parameters from their experiemnt. Smaller plants are more susceptible to ultraviolet radiation than larger plants, as indicated by their LD_{50}'s, which are somewhat similar in magnitude to both infrared and microwave radiation, described earlier.

Andreasen et al. (1999) observed regrowth after irradiation, which suggests that more than one treatment would be necessary to obtain efficient weed control. They also discovered that the distance between the source of UV radiation and the target plants played an important role: increasing the distance from just above the canopy to 17 cm increased the required dose almost twofold (Andreasen et al., 1999).

TABLE 3.7 Summary of Regression Parameters From the Estimated Dose-Response Curves

	Species	Growth Stage	D (g pot^{-1})	C (g pot^{-1})	b	LD$_{50}$ (GJ ha^{-1})	LD$_{50}$ (J cm^{-2})
Weeds	Annual bluegrass	I	2.51	–	0.98	1.23	12.3
		II	11.8	–	0.74	11.1	111.0
	Common groundsel	I	3.32	0.023	1.38	0.50	5.0
		II	17.1	–	0.76	6.22	62.2
	Small nettle	I	2.56	0.03	2.02	0.10	1.0
		II	16.4	6.92	1.34	1.48	14.8
	Shepherd's purse	I	3.18	–	0.67	0.16	1.6
		II	8.94	0.86	0.97	0.5	5.0
Crops	Canola	I	20.7	0.25	1.24	0.75	7.5
	Pea	I	5.45	0.45	1.07	3.13	31.3

Notes: "I" indicates growth of between 2 and 4 leaves; "II" indicates between 20 and 40 leaves, dependent on species.
Modified from Andreasen, C., Leif, H., Jens, C.S., 1999. The effect of ultraviolet radiation on the fresh weight of some weeds and crops. Weed Technol. 13 (3), 554–560.

3.5 LASERS

Light amplification through stimulated emission of radiation (laser) is commonly used for cutting industrial materials, surgery, wood cutting, and research. Laser creates coherent, monochromatic light, which concentrates a large amount of energy into a narrow, nonspreading beam (Heisel et al., 2001). Recently, UV (355 nm), visible (532 nm), IR (810 nm), and CO_2 (1064 nm) lasers have been used to cut the stems of weeds, including perennial ryegrass (*Lolium perenne* L.) (Heisel et al., 2001).

Mathiassen et al. (2006) investigated the effect of laser treatment on common chickweed (*Stellaria media*), scentless mayweed (*Tripleurospermum inodorum*), and canola (*Brassica napus*). Effective treatment requires the laser to be focused onto the apical meristem of the plants (Mathiassen et al., 2006). Several machine vision-based systems have been explored to achieve accurate placement of the laser spot onto weed plants (Mathiassen et al., 2006). Another technique is to move the laser beam back and forth as the system moves forward to achieve good ground coverage. In all cases, it is essential that the laser beam intercepts the weed plant in a favorable way that causes damage to the stem. Proper laser focusing is difficult to achieve in practical terms.

Heisel et al. (2001) found that applying between 0.9 and 2.3 J mm^{-2} from a CO_2 laser, when applied below the meristem, resulted in a 90% or more reduction in weed biomass, in common lamb's quarters (*Chenopodium album*) and wild mustard (*Sinapis arvensis*), respectively. Mathiassen et al. (2006) developed a response equation for laser weed treatment of the form:

$$Y = \frac{D-C}{1+e^{\left[2b\left(\log(\mathrm{LD}_{90})+\frac{1.099}{b}-\log(x)\right)\right]}} + C \tag{3.19}$$

TABLE 3.8 Plant Response to Laser Treatment

Laser	Spot Diameter (mm)	Stellaria media			Tripleurospermum inodorum			Brassica napus		
		b	LD_{90} (J)	LD_{90} (J mm^{-2})	b	LD_{90} (J)	LD_{90} (J mm^{-2})	b	LD_{90} (J)	LD_{90} (J mm^{-2})
5 W, 532 nm	0.9 (0.64 mm^2)	−4.6	1.4	2.2	−3.4	2.6	4.1	n.e.	>5.0	7.8
	1.8 (2.54 mm^2)	n.e.	<1.25	0.5	−5.3	2.7	1.1	−4.7	10.8	4.3
90 W, 810 nm	1.2 (1.13 mm^2)	−3.0	58.3	51.6	−5.4	44.8	39.7	n.e.	>90.0	79.6
	2.4 (4.52 mm^2)	−0.9	104.9	23.2	−3.2	73.8	16.3	−1.6	>225.0	49.8

Note: n.e. indicates that this parameter was unable to be evaluated.

Modified from Mathiassen, S.K., Bak, T., Christensen, S., Kudsk, P., 2006. The effect of laser treatment as a weed control method. Biosyst. Eng. 95 (4), 497–505.

where Y is the fresh weight yield to a UV dose of x (J mm^{-2}), D is the upper limit of fresh weight, C is the lower limit, and b and LD_{90} are determined experimentally. Their results indicate that the efficacy of laser weed control depends on the weed species, wavelength, exposure time, spot size, and laser power (Table 3.8). As with all radiation weed control methods, efficacy increases with power and exposure time; however, in the case of laser-based weed control, spot diameter also affects efficacy. The most efficient system was the 5 W, 532 nm laser with a 1.8 mm spot diameter (Mathiassen et al., 2006).

Most of the values for b in Table 3.8 are fairly large, indicating that the slope of the plant response is high; therefore, it is important to apply an energy dose that is higher than the threshold value for LD_{90}, to ensure efficacy. When the various laser energy doses are scaled in comparison with other weed control techniques, a dose of between 50 and 7960 J cm^{-2} of laser energy is required to control 90% of weeds, depending on the weed species, growth stage, laser power, and laser wavelength, which is actually more than is needed by some other forms of electromagnetic radiation.

Lasers have the potential to provide weed control; however, the device needs to accurately target the weed plant stems to kill the plant. This is not easily achieved. Various machine vision or scanning techniques are being investigated to provide accurate laser targeting for weed control. These systems are also being used in autonomous agricultural robots.

3.6 ROBOTS

One of the first areas of adoption for robotics in agriculture has been the unmanned aerial vehicle (UAV) or drone. Many of these have been used for on farm surveillance, using cameras that operate in both the visible and IR bands of the electromagnetic spectrum. Aerial imagery can also help a farmer better determine which area of a vineyard is ready for harvest (Graves, 2013) or for remote mapping of weeds.

Larger UAV systems have been commercially available for some time. For example, Yamaha makes an unmanned helicopter that can hover over fields and apply pesticides (Graves, 2013). Weighing in at 99 kg and at a total length of 3.63 m and a height of 1.08 m, each helicopter has a load capacity of 28 kg and runs on a two-stroke, horizontally opposed two-cylinder engine (Anon., 2013). Farmers in Japan have used these systems for 20 years or more, to tend rice crops. These helicopters have recently become available in several other countries as well. These systems are very effective at controlling weeds in inaccessible areas; however, they are wirelessly remote controlled rather than being truly autonomous.

On-ground autonomous vehicles have developed from early research into automatic steering systems, which began development in the 1960s (Bechar and Vigneault, 2016). By the 1990s, most agricultural machines were heavy, powerful, high-capacity machines with high operating costs; however, recent research has focused on the development of many small autonomous systems that cooperate with one another via self-deploying network technologies to improve work efficiency, lower costs, and reduced labor requirements (Bechar and Vigneault, 2016).

Multiple cooperating agricultural robotics systems depend on ad hoc wireless data communication networks. Wireless data communication for agricultural purposes relies on one or more sensors of environmental data (machine vision, weed detection, temperature, humidity, etc.), a signal conditioner, an analogue to digital converter, a microprocessor with an external memory chip, and a radio module for wireless communication between other robots and/or base station (Popescu, 2007).

One of the first wireless networks for agriculture was built for the Dutch Lofar Agro research project (Visser, 2005). It consisted of 150 sensor boards (based on the Mica2 Crossbow wireless sensor platform) deployed in a field for gathering temperature and humidity data (Popescu, 2007). The data received by the sensors were handled by a microprocessor implementing the TinyOS operating system and transmitted via radio in the 868/916 MHz band to a field gateway and from there via Wi-Fi to a personal computer for data logging (Popescu, 2007). Since then, there has been considerable interest in using ad hoc wireless networks for automatic data acquisition in agricultural enterprises.

Australia's Commonwealth Scientific and Industrial Research Organization (CSIRO) has developed a series of devices for agricultural wireless network systems. These devices incorporate a Nordic radio with a range of over 1 km that operates on the 433 or 915 MHz band, an integrated solar battery charging circuit, and an extensive range of sensors (Wark et al., 2007). Their system also incorporates a real-time clock chip to reduce microcontroller overheads. Their system was also based on the TinyOS operating system (Wark et al., 2007).

Robotic weed control systems depend on accurate weed identification. Autonomous weed detection, especially in fallowed paddocks, can be achieved using already existing precision agriculture technologies, such as the GreenSeeker technology (Martin et al., 2012). This technology uses normalized difference vegetative index (NDVI), which uses the red, green, and near-infrared bands to assess plants (Martin et al., 2012). It has been adopted to differentiate growing weeds from background stubble and soil in fallowed paddocks.

Langner et al. (2006) introduced the difference index with red threshold (DIRT), for weed detection on mulched croplands. Their investigations into weed detection on mulched areas show that the DIRT performs better than the normalized difference vegetation index (NDVI) for distinguishing between weeds and mulched soil.

The DIRT formula is

$$\text{DIRT} = \frac{\text{sign}(\beta - R) \cdot (\text{NIR} - R)}{(\text{NIR} + R)} \qquad (3.20)$$

where the signum (sign) function returns a positive one (+1) if the argument is positive, returns a zero if the argument is zero, and returns a negative one (−1) if the argument is negative; β is a threshold value that is determined experimentally; R is the reflectance in the red part of the spectrum; and NIR is the reflectance in the NIR part of the spectrum. Langner et al. (2006) used a value of $\beta = 0.12$; however, they suggest that this parameter may range between 0.08 and 0.15.

Several experimental robotic platforms have been developed for agricultural management. For example, Lin et al. (2015) developed a four-wheel drive/four-wheel steering robotic platform for working in wheat cropping systems. Rotation and steering of the wheels are controlled independently using four servomotors for propulsion and four step motors for steering. The platform allows orientation of the vehicle in any desired direction (Lin et al., 2015; Bechar and Vigneault, 2016).

Researchers at Queensland University of Technology have developed small cooperative agricultural robots to increase broadacre crop production and reduce environmental impact. Their project is creating a new class of machines to perform weeding, which is the key element of zero-tillage agriculture (Corke, 2013). These robots have advanced navigation capability using low-cost sensors, unlike current agricultural precision guidance, while also supporting local navigation with respect to weeds and other robots. Their field test exemplify how multiple small robots cause less soil damage, apply herbicide and alternative weed control methods more intelligently, and operate as a system that is more robust to individual machine failures (Corke, 2013). The advent of intelligent delivery platforms, like robots, has allowed several nonrelated technologies to be considered for weed management.

3.7 ABRASIVE WEED CONTROL (WEED BLASTING)

Abrasive grits, which are propelled by compressed air, have been used to remove paint, rust, or grease from hard surfaces (Forcella, 2012). Soft grits, derived from corncobs, nut shells, and seed coats of stone fruits, have been tested for their ability to abrade and control broadleaf and grassweeds. In experiments, grit particles of approximately 0.5 mm diameter were placed in a tank, which was pressurized to between 550 and 700 kPa with compressed air (Forcella, 2012). A high-strength hose connected the tank to a single porcelain nozzle (Forcella, 2012). A single brief exposure (<1 s) to abrasive soft grits typically killed small broadleaf weed seedlings and severely abraded grass seedlings (Forcella, 2012).

Forcella (2013) explored the responses of soybean seedlings to in-row weed abrasion using corncob grit. He applied these weed controls at emergence, cotyledon, unifoliate, first trifoliate, and second trifoliate developmental stages, respectively, in both greenhouse and field settings. Seedling leaf areas and dry weights in the greenhouse experiments were reduced by treatments that included abrasion at cotyledon development, with the primary effect expressed through reductions in the size of the unifoliate leaf. In the field, soybean stand was also reduced by grit applications at the same stage of development, especially if followed by a second application at unifoliate or first trifoliate stage. However, ultimate soybean yield was not reduced by grit application. End-of-season weed dry weights did not differ from hand-weeded controls, and weeds did not impact soybean yields.

In a similar experiment (Forcella, 2012), grit, which was derived from corncobs, was directed at seedlings of summer annual weeds growing at the bases of corn plants when the corn was at differing early stages of leaf development. Season-long, in-row weed control exceeded 90% when two or three abrasion events were coupled with between-row cultivation. Timing of weed abrasion was critical, with highest levels of control corresponding to multiple applications at the one- and five-leaf stages or the one-, three-, and five-leaf stages of corn development. Corn yields associated with these treatments were equivalent to those of hand-weeded controls in which no abrasive grit was applied.

3.8 CONCLUSIONS

All the techniques discussed in this chapter can be used to either control weed growth or kill weeds. In most of the applications of physics to weed control, direct plant damage from physical, electric, or thermal effects causes plant mortality. Many of the techniques discussed in this chapter target already emerged weed plants and in some cases, such as electrocution, laser and abrasive technologies, and ancillary targeting system, such as machine vision or robotics, are required for good efficacy.

At least two of these physical weed control technologies, namely, steam and microwave treatment, also offer opportunity to deactivate weed seeds in the soil seed bank. Both technologies rely on a thermal deactivate of dormant seeds. Most of the technologies discussed in this chapter require moderate to high energy investment, and in some cases, such as electrocution or electrostatic fields, there are high human health and safety risks to be considered as well. Some of these technologies have been commercialized to some degree; however, many of them have not progressed beyond the research phase. As herbicide resistance becomes more prevalent, some or all of these technologies may become more widely adopted.

APPENDIX A

Determining the peak radiation wavelength for any temperature:

$$\frac{d\rho}{d\lambda} = \frac{2c^3 h^2 e^{\frac{hc}{\lambda kT}}}{kT\lambda^7 \left(e^{\frac{hc}{\lambda kT}} - 1\right)^2} - \frac{10c^2 h}{\lambda^6 \left(e^{\frac{hc}{\lambda kT}} - 1\right)} = 0$$

$$\frac{2c^3 h^2 e^{\frac{hc}{\lambda kT}}}{kT\lambda^7 \left(e^{\frac{hc}{\lambda kT}} - 1\right)^2} = \frac{10c^2 h}{\lambda^6 \left(e^{\frac{hc}{\lambda kT}} - 1\right)}$$

Rearranging gives

$$\lambda = \frac{2h^2c^3 e^{\frac{hc}{\lambda kT}}\left(e^{\frac{hc}{\lambda kT}} - 1\right)}{10c^2hkT\left(e^{\frac{hc}{\lambda kT}} - 1\right)^2}$$

or

$$\lambda = \frac{hc \cdot e^{\frac{hc}{\lambda kT}}}{5kT\left(e^{\frac{hc}{\lambda kT}} - 1\right)}$$

At normal temperatures $e^{\frac{hc}{\lambda kT}} \sim 10^{20}$, therefore,

$$\lambda \approx \frac{hc}{5kT}$$

APPENDIX B

Determining the total radiated power from a black body,

$$P = 2hc^2 \int_0^\infty \frac{1}{\lambda^5 \left\{e^{\frac{hc}{\lambda kT}} - 1\right\}} \cdot d\lambda$$

Let

$$u = \frac{hc}{\lambda kT}; \ \lambda = \frac{hc}{ukT} \ \text{ and } \ d\lambda = \frac{-hc}{u^2 kT} \cdot du$$

Substituting into the previous equation gives

$$P = 2hc^2 \int_0^\infty \frac{\frac{-hc}{u^2 kT}}{\left(\frac{hc}{ukT}\right)^5 \{e^u - 1\}} \cdot du$$

Rearranging gives

$$P = 2hc^2 \left(\frac{kT}{hc}\right)^4 \int_0^\infty \frac{u^3}{\{e^u - 1\}} \cdot du$$

Evaluating the integral gives

$$P = 2hc^2 \left(\frac{kT}{hc}\right)^4 \frac{\pi^4}{15}$$

Rearranging gives

$$P = \frac{2\pi^5 k^4}{15h^3 c^2} T^4$$

This can be simplified to

$$P = \sigma T^4$$

where σ is the referred to as the Stefan-Boltzmann constant.

In the case of a normal object, the power transfer is reduced by a factor ε, depending on the properties of the object's surface; therefore, the power transfer is

$$P = \varepsilon \sigma T^4$$

References

Andreasen, C., Leif, H., Jens, C.S., 1999. The effect of ultraviolet radiation on the fresh weight of some weeds and crops. Weed Technol. 13 (3), 554–560.

Anon., 2013. Yamaha Unveils Unmanned Helicopters. Available from, https://www.tradefarmmachinery.com.au/product-news/1307/yamaha-unveils-unmanned-helicopters.

Ark, P.A., Parry, W., 1940. Application of high-frequency electrostatic fields in agriculture. Q. Rev. Biol. 15 (2), 172–191.

Ascard, J., 1994. Dose-response models for flame weeding in relation to plant size and density. Weed Res. 34 (5), 377–385.

Ascard, J., 1998. Comparison of flaming and infrared radiation techniques for thermal weed control. Weed Res. 38, 69–76.

Bachman, C.H., Reichmanis, M., 1973. Barley leaf tip damage resulting from exposure to high electrical fields. Int. J. Biometeorol. 17 (3), 243–251.

Barker, A.V., Craker, L.E., 1991. Inhibition of weed seed germination by microwaves. Agron. J. 83 (2), 302–305.

Bebawi, F.F., Cooper, A.P., Brodie, G.I., Madigan, B.A., Vitelli, J.S., Worsley, K.J., Davis, K.M., 2007. Effect of microwave radiation on seed mortality of rubber vine (*Cryptostegia grandiflora* R.Br.), parthenium (*Parthenium hysterophorous* L.) and bellyache bush (*Jatropha gossypiifolia* L.). Plant Prot. Q. 22 (4), 136–142.

Bechar, A., Vigneault, C., 2016. Agricultural robots for field operations: concepts and components. Biosyst. Eng. 149, 94–111.

Brodie, G., 2007. Simultaneous heat and moisture diffusion during microwave heating of moist wood. Appl. Eng. Agric. 23 (2), 179–187.

Brodie, G., 2008. The influence of load geometry on temperature distribution during microwave heating. Trans. ASABE 51 (4), 1401–1413.

Brodie, G., Botta, C., Woodworth, J., 2007a. Preliminary investigation into microwave soil pasteurization using wheat as a test species. Plant Prot. Q. 22 (2), 72–75.

Brodie, G., Hamilton, S., Woodworth, J., 2007b. An assessment of microwave soil pasteurization for killing seeds and weeds. Plant Prot. Q. 22 (4), 143–149.

Brodie, G., Harris, G., Pasma, L., Travers, A., Leyson, D., Lancaster, C., Woodworth, J., 2009. Microwave soil heating for controlling ryegrass seed germination. Trans. ASABE 52 (1), 295–302.

Brodie, G., Hollins, E., 2015. The effect of microwave treatment on ryegrass and wild radish plants and seeds. Glob. J. Agric. Innov. Res. Dev. 2 (1), 16–24.

Brodie, G., Pasma, L., Bennett, H., Harris, G., Woodworth, J., 2007c. Evaluation of microwave soil pasteurization for controlling germination of perennial ryegrass (*Lolium perenne*) seeds. Plant Prot. Q. 22 (4), 150–154.

Brodie, G., Ryan, C., Lancaster, C., 2012. The effect of microwave radiation on paddy melon (*Cucumis myriocarpus*). Int. J. Agron. 2012, 1–10.

Chappell, J. M., Iqbal, A., Abbott, D., 2010. A simplified approach to electromagnetism using geometric algebra. Phys. Educ., Available from: https://arxiv.org/pdf/1010.4947.pdf.

Clark, W.J., Kissell, C.W., 2003. System and Method for In Situ Soil Sterilization, Insect Extermination and Weed Killing. Patent No. 20030215354A1.

Corke, P., 2013. Robotics for Zero-Tillage Agriculture. Available from https://wiki.qut.edu.au/display/cyphy/Robotics+for+zero-tillage+agriculture.

Davis, F.S., Wayland, J.R., Merkle, M.G., 1971. Ultrahigh-frequency electromagnetic fields for weed control: phytotoxicity and selectivity. Science 173 (3996), 535–537.

Davis, F.S., Wayland, J.R., Merkle, M.G., 1973. Phytotoxicity of a UHF electromagnetic field. Nature 241 (5387), 291–292.

Debye, P., 1929. Polar Molecules. Chemical Catalog, New York.

Diprose, M.F., Benson, F.A., Hackam, R., 1980. Electrothermal control of weed beet and bolting sugar beet. Elektrothermische Bekämpfung von Unkrautrüben und Zucker-rübenschossern 20 (5), 311–322.

Diprose, M.F., Benson, F.A., Willis, A.J., 1984. The effect of externally applied electrostatic fields, microwave radiation and electric currents on plants and other organisms, with special reference to weed control. Bot. Rev. 50 (2), 171–223.

Dirac, P.A.M., 1927. The quantum theory of the emission and absorption of radiation. Proc. R Soc. Lond A Math. Phys. Sci. 114 (767), 243–265.

Duke, S.O., 2010. Herbicide and pharmaceutical relationships. Weed Sci. 58 (3), 334–339.

Einstein, A., 1920. Relativity: The Special and General Theory (R.W. Lawson, Trans.). Henry Holt and Company, New York (Original work published 1916).

Einstein, A., 1951. The advent of the quantum theory. Science 113 (2926), 82–84.

Evans, M.W., 2005. The spinning and curving of spacetime: the electromagnetic and gravitational fields in the Evans field theory. Found. Phys. Lett. 18 (5), 431–454.

Forcella, F., 2012. Air-propelled abrasive grit for postemergence in-row weed control in field corn. Weed Technol. 26 (1), 161–164.

Forcella, F., 2013. Soybean seedlings tolerate abrasion from air-propelled grit. Weed Technol. 27 (3), 631–635.

Gay, P., Piccarolo, P., Ricauda Aimonino, D., Tortia, C., 2010a. A high efficacy steam soil disinfestation system, part II: design and testing. Biosyst. Eng. 107 (3), 194–201.

Gay, P., Piccarolo, P., Ricauda Aimonino, D., Tortia, C., 2010b. A high efficiency steam soil disinfestation system, part I: physical background and steam supply optimisation. Biosyst. Eng. 107 (2), 74–85.

Giancoli, D.C., 1989. Physics for Scientists and Engineers, second ed. Prentice Hall, New Jersey.

Gibson, B.F., Frances, M.F., Deacon, J.W., 1988. Effects of microwave treatment of soil on growth of birch (*Betula pendula*) seedlings and infection of them by ectomycorrhizal fungi. New Phytol. 108, 189–204.

Gourd, T., 2002. Controlling Weeds Using Propane Generated Flame and Steam Treatments in Crop and Non Croplands. Organic Farming Research Foundation, Santa Cruz, CA.

Graves, B., 2013. Growing reliance on technology: agriculture has high-tech helpers. San Diego Bus. J. 34 (43), 1.

Grigorov, G.R., 2003. Method and System for Exterminating Pests, Weeds and Pathogens. Patent No. 20030037482A1.

Guyton, K.Z., Loomis, D., Grosse, Y., El Ghissassi, F., Benbrahim-Tallaa, L., Guha, N., Scoccianti, C., Mattock, H., Straif, K., 2015. Carcinogenicity of tetrachlorvinphos, parathion, malathion, diazinon, and glyphosate. Lancet Oncol. 16 (5), 490–491.

Haller, H.E., 2002. Microwave Energy Applicator. Patent No. 20020090268A1.

Heap, I.M., 1997. The occurrence of herbicide-resistant weeds worldwide. Pestic. Sci. 51 (3), 235–243.

Heap, I.M., 2008. International Survey of Herbicide Resistant Weeds. Available from, http://www.weedscience.org/in.asp.

Heisel, T., Schou, J., Christensen, S., Anderson, C., 2001. Cutting weeds with a CO_2 laser. Weed Res. 41, 19–29.

Hendrick, J., 1918. Experiments on the treatment of growing crops with overhead electric discharges. Scott. J. Agric. 1, 160–171.

Henry, P.S.H., 1948. The diffusion of moisture and heat through textiles. Discuss. Faraday Soc. 3, 243–257.

Hernández, A.F., Parrón, T., Tsatsakis, A.M., Requena, M., Alarcón, R., López-Guarnido, O., 2013. Toxic effects of pesticide mixtures at a molecular level: their relevance to human health. Toxicology 307, 136–145.

Holman, J.P., 1997. Heat Transfer, eighth ed. McGraw-Hill, New York.

International Telecommunication Union, 2004. Spectrum Management for a Converging World: Case Study on Australia. International Telecommunication Union.

Jorgensen, I., Stiles, W., 1917. Atmospheric electricity as an environmental factor. J. Ecol. 5, 203–209.

Kelley, M.C., 2013. The Earth's Electric Field. [Electronic Resource]: Sources from Sun to Mud. 2013 Elsevier Science, Burlington.

Langner, H.-R., Böttger, H., Schmidt, H., 2006. A special vegetation index for the weed detection in sensor based precision agriculture. Environ. Monit. Assess. 117 (1), 505–518.

Lepeschkin, W.W., 1912. Zur Kenntnis der Einwirkung supamaximaler Temperaturen auf die Pflanze. Ber. Deut. Bot. Ges. 30, 713–714.

Levitt, J., 1980. Response of Plants to Environmental Stresses. vol. 1. Academic Press, New York.

Lin, H., Dong, S., Liu, Z., Yi, C., 2015. Study and experience on a wheat precision seeding robot. J. Robot. 696301.

Mačkić, S., Ahmetović, N., 2011. Toxicological profiles of highly hazardous herbicides with special reference to carcinogenicity to humans. Herbologia 12 (2), 55–60.

Martin, D.E., López, J.D., Lan, Y., 2012. Laboratory evaluation of the GreenSeeker handheld optical sensor to variations in orientation and height above canopy. IJABE 5 (1), 43–47.

Mathiassen, S.K., Bak, T., Christensen, S., Kudsk, P., 2006. The effect of laser treatment as a weed control method. Biosyst. Eng. 95 (4), 497–505.

Meissner, T., Wentz, F.J., 2004. The complex dielectric constant of pure and sea water from microwave satellite observations. IEEE Trans. Geosci. Remote Sens. 42 (9), 1836–1849.

Menges, R.M., Wayland, J.R., 1974. UHF electromagnetic energy for weed control in vegetables. Weed Sci. 22 (6), 584–590.

Metaxas, A.C., Meredith, R.J., 1983. Industrial Microwave Heating. Peter Peregrinus, London.

Montoro, T., Manrique, E., Gonzalez-Reviriego, A., 1999. Measurement of the refracting index of wood for microwave radiation. Holz Roh Werkst. 57 (4), 295–299.

Murr, L.E., 1963a. Optical microscopy investigation of plant cell destruction in an electrostatic field. J. Pa. Acad. Sci. 37, 109–121.

Murr, L.E., 1963b. Plant growth response in a simulated electric field environment. Nature 200, 490–491.

Nelson, S.O., 1996. A review and assessment of microwave energy for soil treatment to control pests. Trans. ASAE 39 (1), 281–289.

Nelson, S.O., Stetson, L.E., 1985. Germination responses of selected plant species to RF electrical seed treatment. Trans. ASAE 28 (6), 2051–2058.

Newman, J.E., 1911. Electricity as applied to agriculture. Electrician, 915. March 17th.

Owen, M., Walsh, M., Llewellyn, R., Powles, S., 2007. Widespread occurrence of multiple herbicide resistance in western Australian annual ryegrass (Lolium rigidum) populations. Aust. J. Agric. Res. 58 (7), 711–718.

Parish, S., 1990. A review of non-chemical weed control techniques. Biol. Agric. Hortic. 7 (2), 177–1137.

Peighambarzadeh, S.Z., Safi, S., Shahtaheri, S.J., Javanbakht, M., Forushani, A.R., 2011. Presence of atrazine in the biological samples of cattle and its consequence adversity in human health. Iran. J. Public Health 40 (4), 112–121.

Popescu, V., 2007. Wireless data communication in agricultural engineering. Trends and practical experiments.Proc. Research People and Actual Tasks on Multidisciplinary Sciences. Lozenec, Bulgaria.

Raffaelli, M., Martelloni, L., Frasconi, C., Fontanelli, M., Carlesi, S., Peruzzi, A., 2016. A prototype band-steaming machine: design and field application. Biosyst. Eng. 144, 61–71.

Sartorato, I., Zanin, G., Baldoin, C., De Zanche, C., 2006. Observations on the potential of microwaves for weed control. Weed Res. 46 (1), 1–9.

Serdyuk, V.M., 2008. Dielectric properties of bound water in grain at radio and microwave frequencies. Prog. Electromagn. Res. 84, 379–406.

Shibusawa, M., Shibata, K., 1930. The effect of electric discharges on the rates of growth of plants. Abstract No. 23923, Biol. Abstr. 4 (10), 2257.

Singh, R.P., Heldman, D.R., 1993. Introduction to Food Engineering, second ed. Academic Press, New York.

Smith, R.J., 1976. Circuits, Devices and Systems, third ed. Wiley International, New York.

Speir, T.W., Cowling, J.C., Sparling, G.P., West, A.W., Corderoy, D.M., 1986. Effects of microwave radiation on the microbial biomass, phosphatase activity and levels of extractable N and P in a low fertility soil under pasture. Soil Biol. Biochem. 18 (4), 377–382.

Storeheier, K., 1994. Basic investigations into flaming for weed control. Acta Hortic. 372, 195–204.

Tran, V.N., 1979. Effects of microwave energy on the strophiole, seed coat and germination of acacia seeds. Aust. J. Plant Physiol. 6 (3), 277–287.

Troudi, A., Sefi, M., Ben Amara, I., Soudani, N., Hakim, A., Zeghal, K.M., Boudawara, T., Zeghal, N., 2012. Oxidative damage in bone and erythrocytes of suckling rats exposed to 2,4-dichlorophenoxyacetic acid. Pestic. Biochem. Physiol. 104 (1), 19–27.

Ulaby, F.T., El-Rayes, M.A., 1987. Microwave dielectric spectrum of vegetation—part II: dual-dispersion model. IEEE Trans. Geosci. Remote Sens. GE-25 (5), 550–557.

Visser, O.W., 2005. Localisation in Large-Scale Outdoor Wireless Sensor Networks (Unpublished thesis). Delft University of Technology, Faculty of Electrical Engineering, Mathematics, and Computer Science.

Vollmer, M., 2004. Physics of the microwave oven. Phys. Educ. 39 (1), 74–81.

Wark, T., Corke, P., Sikka, P., Klingbeil, L., Ying, G., Crossman, C., Valencia, P., Swain, D., Bishop-Hurley, G., 2007. Transforming agriculture through pervasive wireless sensor networks. IEEE Pervasive Comput. 6 (2), 50–57.

Wayland, J.R., Davis, F.S., Merkle, M.G., 1973. Toxicity of an UHF device to plant seeds in soil. Weed Sci. 21 (3), 161.

Welty, J.R., Wicks, C.E., Wilson, R.E., Rorrer, G.L., 2007. Fundamentals of Momentum, Heat and Mass Transfer, fifth ed. John Wiley and Sons, Hoboken, NJ.

Wickerham, E.L., Lozoff, B., Shao, J., Kaciroti, N., Xia, Y., Meeker, J.D., 2012. Reduced birth weight in relation to pesticide mixtures detected in cord blood of full-term infants. Environ. Int. 47, 80–85.

Wolf, W.W., Vaughn, C.R., Harris, R., Loper, G.M., 1993. Insect radar cross-section for aerial density measurement and target classification. Trans. ASABE 36 (3), 949–954.

Weed Control Using Ground Cover Systems

Khawar Jabran*, Bhagirath S. Chauhan†

*Düzce University, Düzce, Turkey †The University of Queensland, Gatton, QLD, Australia

4.1 INTRODUCTION

Weeds are a competitor to plants in all kinds of farming. Non-chemical weed control is desired due to several reasons (Chauhan and Gill, 2014). Most important among these are the demand for pesticide-free foods, evolution of herbicide resistance in weeds, and environmental and health problems caused by herbicides. A number of non-chemical weed control techniques are under investigation by the researchers in all parts of the world. Any non-chemical weed control method may be preferred over the others owing to factors such as nature of crop, ecological characteristics of the area, nature and intensity of weeds, availability and effectiveness of other weed control methods, and social and economic factors (Hatcher and Melander, 2003; Olson and Eidman, 1992). Generally, there is a demand for pesticide-free foods by the consumers, which is endorsed by the scientific evidence proving pesticide-free food beneficial over the pesticide-contaminated food (Bourn and Prescott, 2002; Crinnion, 2010; Lairon, 2011; Winter and Davis, 2006). Importantly, the vegetables and field fruit crops should be particularly pesticide-free. Vegetables and field fruits generally stay for lesser time in the field than the field crops; this period may be insufficient for the loss or decay of herbicides (applied for weed control) received by these plants. Hence, there is greater likeliness that vegetables and field fruits contain a higher herbicide residue than the cereal and other field crops (Lu et al., 2006). This makes non-chemical weed control more important for vegetables and field fruits than the regular field crops.

The results of recent research indicate that ground covers are highly useful for controlling weeds in vegetables and field fruits and provide several additional benefits (such as soil and water conservation) (Abdullah et al., 2015; Jabran et al., 2015a,c). Covering the ground with some stuff can inhibit the penetration of light to weed seeds (Mo et al., 2016). Also, this can exert a physical pressure on weeds to suppress their growth. Such coverings may include a plastic sheet, straw mulch, woodchips, carpets, or paper mulches (Hammermeister, 2016). A slight rise in soil temperature, a blockage of sunlight to reach the weed seeds, and a physical pressure on emerging weeds are the probable mechanisms through which mulches suppress the weeds (Haapala et al., 2015). The nondegradable residue of plastic covering and slow decomposition of straw mulch are the important issues in using these mulches for weed control. Use of degradable mulches has been suggested as a solution to such problems (Hammermeister, 2016).

The objective of this chapter is to discuss the use of ground-cover system (GCS) for non-chemical weed control in different agricultural systems. The GCS techniques discussed in this chapter include the plastic covering, straw mulch covering, woodchips, and the use of mats and carpets for weed control. The chapter also discusses the need for the use of degradable/decomposable mulches.

4.2 PLASTIC (BLACK POLYETHYLENE) COVERING

Use of plastic for covering the soil is gaining popularity in many agricultural systems. This provides several benefits that are not limited to soil conservation, improved water productivity, an increase in economic benefits, regulation of soil temperature, and weed control (Chang et al., 2016; Jabran et al., 2016; Lament, 1993). Only a black (or occasionally a colored one) plastic mulch is recommended by researchers for use in agricultural systems as transparent plastic mulch is not as effective as the black one; the lack of blockage of sunlight may be the most probable reason for that (Chang et al., 2016). Under the plastic ground covering, the productivity of many crops, particularly vegetables (sweet corn, potato, muskmelon, pepper, okra, tomato, squash, eggplant, etc.), is known to increase significantly (Chang et al., 2016; Lament, 1993).

Black plastic mulch strongly inhibits weeds by exerting a physical pressure on these, blocks the sunlight to reach the weeds or weed seeds (hence inhibiting the weed germination), and warms the soil causing a solarization impact (Chang et al., 2016; Mo et al., 2016). These all cause damage to weed seeds, inhibit weed seed germination, harm the germinating weeds, and suppress or at least decrease the growth of established weeds (Chang et al., 2016; Chauhan, 2013). Use of a mulch covering may also cause an oxygen stress owing to oxygen depletion in the soil that will negatively affect the germination or growth of weeds (Boyd and Van Acker, 2004).

Solarization is important among the impacts of plastic mulch that makes it important in suppressing weeds (Khan et al., 2012). Under solarization, the black plastic mulch entraps the radiations from the sun to heat the soil (Katan, 2015). The temperature of soil rises several degrees over the normal soil temperature. For instance, the soil temperature may reach ~50°C (Katan, 2015). This causes the death of sprouts, damages weed seeds, and inhibits weed seed germination (Abouziena and Haggag, 2016). Soil solarization has been discussed in detail in Chapter 2 of this book.

A rational productivity of watermelon and weed control was achieved through the use of plastic covering in Brazil (da Silva et al., 2013). A study from the United States indicated that the black plastic mulch decreased weed biomass by 90% (Splawski et al., 2016). Similarly, the use of plastic mulch not only decreased the weed infestation in tomato and strawberry production but also increased their yield significantly over the control treatment (Johnson and Fennimore, 2005; Radics and Szné Bognár, 2002). In a study from Egypt, the plastic mulch was found highly effective in controlling weeds and economizing the use of water (Hegazi, 2000). In Vietnam, the black plastic mulch proved highly effective for control of several weeds growing in peanut; the yield of the mulched plots was also increased significantly over the control treatment (Ramakrishna et al., 2006). In Pakistan, the use of black polyethylene not only helped to suppress weeds but also achieved the highest tomato yield over the control and other treatments that included various kinds of mulches and herbicide application (Bakht and Khan, 2014).

Black polyethylene mulching has been found to provide effective weed suppression in many field crops in addition to vegetables. For instance, this mulching could effectively decrease the density and biomass of weeds in maize and increase maize productivity (higher than hand weeding and a herbicidal weed control) (Khan et al., 2016). Black polyethylene provided encouraging results when used for controlling weeds in direct-seeded rice in Pakistan (personnel observation; data not published). However, the method and time of the application of black plastic mulch in direct-seeded rice (relevant to weed emergence) have been important factors in achieving the desired results. Applying the plastic mulch between the rice rows and after weed emergence may not provide an effective weed control. The most appropriate way is to spread the plastic sheet in the whole field and provide holes (in sheet) only for crop plant emergence. Fertilizer management may be a difficult task when the fields are fully covered with plastic sheets. This may be easy when plastic covering is combined with drip irrigation. Such management will help both in water saving and weed control.

The fruit plants and forests may also benefit from the positive impacts (including weed control) of covering the soil with a plastic mulch (Hammermeister, 2016). For example, a black plastic covering could improve the moisture retention and suppress weeds in Chinese date (*Ziziphus mauritiana*) (Singh and Ghosal, 2016).

In addition to several benefits, some drawbacks of using plastic mulch have also been noted. The high cost of the plastic mulch compared with other methods of weed control (e.g., the use of herbicides) is one among the demerits of the plastic mulch (Hegazi, 2000). Although effective in suppressing weeds, yet the plastic mulch may require a huge number of labor to collect its leftovers after the crop season ends (Schonbeck, 1999). Nutrient management may become a difficult task under plastic mulching, particularly in case of fertilizers that are applied post plastic mulch application (Hammermeister, 2016). Another important disadvantage of the plastic mulch includes the environmental pollution caused by its remains/residues after the cropping season is over. The plastic mulch is nondegradable; its residues in the field not only pollute the environment but also cause hindrances in farm operations (Jiang et al., 2017). For instance, Steinmetz et al. (2016) have described the negative effects of plastic mulching on soil environment. The chemical part of plastic may become intact in the soil to cause certain negative impacts and lower the soil quality. Also, there is likeliness for an enhanced carbon and nitrogen emissions from the soil under the plastic mulch (Steinmetz et al., 2016). In a study from China, plastic mulching resulted in increased CO_2 emissions from the soil; however, it reduced the overall intensity of greenhouse gases from wheat and maize fields (Chen et al., 2016). Alternatives to the polyethylene black mulch (such as a biodegradable mulch) are required to benefit from this excellent technique of weed control (Hammermeister, 2016; Moreno and Moreno, 2008).

4.3 STRAW COVERING

For most of the field crops, the grains are utilized as food, while straw is fed to animals, burnt, used for producing several products (e.g., biogas), or retained in the field. Retention of crop residues in the field has been included in the principles of conservation agriculture (Farooq et al., 2011a). The aim is to conserve the soil and water and suppress weeds (Abouziena and Haggag, 2016; Erenstein, 2003; Farooq et al., 2011a,b; Jabran et al., 2015a). Researchers strongly recommend the retention or application of straw covering over the soil for attaining benefits such as increased soil fertility, enhanced soil moisture retention, improved grain quality, and weed control (Table 4.1) (Jabran et al., 2016, 2015b,c; Ramakrishna et al., 2006). Retention of straw may effectively suppress weeds in the following crop. For instance, the wheat crop in rice-wheat system of South Asia usually has a high weed infestation (Jabran et al., 2012; Razzaq et al., 2012). Rice residues left in the field may help to suppress weeds in wheat, particularly, when the crop is sown with a zero-tillage drill (Rahman et al., 2005). The drill removes the straw from the line where wheat seeds are sown, while the rest of the soil surface gets a cover of rice straw. In the same cropping system (i.e., rice-wheat), the straw from wheat may also be retained to control weeds in rice (Singh et al., 2007). Use of wheat straw as mulch at 4 t/ha was helpful in suppressing weeds in direct-seeded rice (Singh et al., 2007). A study from Germany indicated that straw (4 or 6 t/ha) retained from previous oat crop could suppress the weeds (*Chenopodium album* L. and *Matricaria* spp.) in the following organically grown faba bean (Massucati and Kopke, 2014). Perennial weeds

TABLE 4.1   Different Straw Coverings Used for Weed Control in Different Parts of the World

Straw Covering	Rate	Cultivated Crop	Region	References
Rye	–	Soybean	The United States	Liebl et al. (1992)
Oat	4–6 t/ha	Faba bean	Germany	Massucati and Kopke (2014)
Rye Buckwheat	10 and 20 t/ha	Broccoli	Poland	Kosterna (2014)
Corn Rape	10 and 20 t/ha	Tomato	Poland	Kosterna (2014)
Rice straw Barley straw Maize harvest residue	1 kg/m^2	Tomato	Spain	Anzalone et al. (2010)
Wheat straw	1 kg/m^2	Tomato	Pakistan	Bakht and Khan (2014)
Wheat straw	4 t/ha	Rice	India	Singh et al. (2007)
Rice straw	6 t/ha	Ginger	India	Thankamani et al. (2016)
Rice straw	10 t/ha	Groundnut	Vietnam	Ramakrishna et al. (2006)

and grasses however were difficult to be controlled through the use of retained straw (Massucati and Kopke, 2014). Under such instances, that is, where a few weeds are not controlled through the use of a certain GCS, the employed technique may be aided with other weed control methods (e.g., mechanical weed control); hence, effective weed control may be achieved through the use of integrated weed management. An allelopathic effect of certain straw mulches may also be noted on weeds (Farooq et al., 2011b; Jabran and Farooq, 2013); however, this chapter does not address the allelopathic weed control. The use of allelopathy for weed control has been provided in the fifth and sixth chapters of this book.

The straw covering may exert multiple effects in suppressing weeds. This may include a physical pressure of straw on emerging weeds, blockage of sunlight, and exudation of allelochemicals (if the source of mulch is an allelopathic plant) (Jabran and Farooq, 2013; Singh et al., 2007). Other than this, mulch can also warm the soil temperature that may cause the death of germinating weed seedlings. The crop may also face some negative effects from the straw mulch if crop seeds or seedlings are covered by the mulch. This can be avoided by leaving the crop rows free of straw mulch cover. The effectiveness of straw mulch has been known since long for weed suppression in agricultural systems (Crutchfield et al., 1986). Recent literature also supports the idea to use straw mulch for weed control. For instances, rice straw was useful in controlling weeds in the wheat crop (Chaudhary and Iqbal, 2013). In a study from Malaysia, the use of rice straw in the form of a mat provided highly effective results for controlling weeds in rice grown by the system of rice intensification (Mohammed et al., 2016). This discussion concludes that the straw mulches can suppress the weeds in various cropping systems.

4.4 PAPER/NEWSPAPER/CLOTH MULCH

Paper leaves or newspapers can be used for controlling weeds in vegetable gardens. In order to achieve a good weed control, the soil being applied with a paper mulch should be free from weeds and be applied with multiple layers (e.g., 2–4 layers) of paper/newspaper. That means the land is first made free from the prevailing weeds and then applied with paper mulch. Recently, a review by Haapala et al. (2014) provides usefulness of paper mulches in various agricultural systems. In contrast to polyethylene mulch, the paper mulches are degradable and environment friendly. The best time to apply paper mulch is before or immediately after the germination of weeds. Paper mulch will inhibit the germination, emergence, and growth of weeds by blocking sunlight and exerting a physical pressure on emerging weed seedlings. Recent studies provide evidence regarding the usefulness of paper mulch in suppressing weeds (Anzalone et al., 2010; Radics and Szné Bognár, 2002). For example, paper mulch helped to achieve effective weed control and increased yield over the control treatment in tomato and green bean (Radics and Szné Bognár, 2002). Newspaper mulch had high effectiveness in decreasing the weed dry weight when applied in pumpkin crop in the United States (Splawski et al., 2016). A study from Spain indicated that brown kraft paper helped to effectively suppress weeds such as *Digitaria sanguinalis* (L.) Scop., *Cyperus rotundus* L., *C. album* L., and *Portulaca oleracea* L. in tomato (Anzalone et al., 2010). Use of kraft paper for weed control has also been reported from the United States; however, it was degraded during the crop season (Schonbeck, 1999). Although paper mulch provided a moderate- or low-level control of weeds in tomato, yet it helped to cause a great increase in the tomato yield over the control treatment (Bakht and Khan, 2014).

4.5 WOOD CHIPS

Woodchips can be used to cover the soil aiming for weed inhibition (Splawski et al., 2016). Flower and vegetable gardens can be affixed with woodchips in order to prevent weed emergence or suppress the growth of emerging weeds (Ingels et al., 2012). A wood layer of 4–6 cm over the soil surface can help to achieve weed suppression. Nevertheless, the woodchip ground covering will receive damages over the time owing to environmental and other factors; hence, a maintenance or rebuild of the woodchip covering will be required (Hammermeister, 2016). A woodchip mulch should be established over a seedbed that is currently free from weed infestation. Any weeds prevailing can be cleared at the time of woodchip covering establishment. Use of woodchips as a ground cover may not provide 100% weed control; the emerging weeds will be required to be managed by other methods. Occasionally, emerging weeds can be pulled out through hand weeding. Sometimes, supplemental weed control may be required if there is high weed emergence, or, in order to fix weeds emerging in areas in the field that are not covered by woodchips.

A study from the United States indicated that woodchips were helpful in suppressing weeds that were growing in pumpkin (Splawski et al., 2016). Woodchips were also helpful in controlling weeds in organically grown pear (Ingels et al., 2012). Wang et al. (2012) from Germany evaluated the efficacy of woodchips in organically grown lentil or lentil/barley intercropping. Woodchips contained chopped and mixed hedgerow shrubs at a rate of 160 m^3/ha. The woodchip mulching was helpful in reducing weeds in either of the cropping; however, it did not increase the grain yield of any crops (Wang et al., 2012). The authors suggested the use of these woodchips for weed control in situations where the use of herbicides is prohibited and mechanical weed control is difficult to be accomplished.

4.6 USE OF MATS AND CARPETS FOR WEED CONTROL

Weed control mats are produced from different kinds of fabric for controlling weeds in organic vegetable production, forest seedlings, and other agricultural situations where herbicide application is not appropriate. Mechanism of weed control by mats includes physical suppression and blockage of sunlight. Weed mats are usually permeable (allowing easy infiltration of water) and degradable (George and Brennan, 2002). Other than this, the carpets that become old and have no utility can be used to cover the weeds. This will result in the killing of emerged and germinating weeds and sprouts through physical pressure of carpet and blockage of sunlight. Blockage of sunlight to the soil surface will inhibit the germination of weed seeds. A carpet being used for weed control should be permeable to allow water infiltration. A list of possible soil covering in the form of mats or carpets, etc. has been provided in Table 4.2.

In a study from the United States, the use of weed mat offered cost-effective weed control in organically produced blackberries (Harkins et al., 2014). The production of blackberries was

TABLE 4.2 Different Kinds of Natural and Synthetic Mats/Carpets Mulches Usable for Natural Weed Control

Mulch Material	Explanation	Reference
Coir fiber mat	Made of coir fiber and backed by photodegradable membrane	Stokes (2012)
Landscape fabric	Manufactured fabric for weed control	Ingels et al. (2012)
Acetylated kenaf	Made of acetylated kenaf (80%) and cotton (20%)	Haywood (1999)
Cardboard	Recycled cardboard	Haywood (1999)
Cellulose mat	Fiber mat of secondary cellulose	Haywood (1999)
Cotton shoddy	Nonwoven cotton	Haywood (1999)
Cotton-polyethylene	Nonwoven cotton backed by polyethylene	Haywood (1999)
Dahoma	Fiber (90%) of dahoma (*Piptadeniastrum africanum*) plus cotton (10%)	Haywood (1999)
Pine straw	Dried pine leaves used at 4.5 kg/m^2	Haywood (1999)
Sunbelt	Polypropylene mat, black in color	Haywood (1999)
Terra Mat. "P"	Polyester mat	Haywood (1999)
Thermat	Plastic laminate mat having one green and other brown side	Haywood (1999)
Weed Barrier	Polypropylene mat	Haywood (1999)
Weed Block 6+	Black polyethylene sheet having several perforations	Haywood (1999)

increased to double by the use of weed mat over the unweeded control (Harkins et al., 2014). Mulching with mats may provide soil protection and improved water retention and weed control during the establishment phase of a forest. For instance, jute mat was used in Australia for controlling weeds in newly sown seedlings of dunns white gum (*Eucalyptus dunnii*) and Sydney blue gum (*E. saligna*) (George and Brennan, 2002). Jute map suppressed weeds and increased the growth of seedling by ∼200% over the control until 2 years; however, the jute degraded later on providing a chance to weeds to grow profusely. Moreover, weed control using jute mat was costlier than herbicide (George and Brennan, 2002). Another study in the United States used cotton and polyester mats (among the other treatments) in order to check weeds in longleaf pine (*Pinus palustris* Mill.) during the seedling stage (Haywood, 2000). These were helpful in improving the growth of seedlings and suppressing weeds for several seasons (Haywood, 2000). A study from England indicated that a mat made from coir fiber was most appropriate for long-term weed suppression and improved growth of common ash (*Fraxinus excelsior* L.) during the establishment stage (Stokes, 2012). Several of the other biodegradable mulches could also suppress weeds but were unacceptable due to certain reasons. For instance, there was a high cost for rubber mats, hardboards were difficult in use, and polyethylene left its residues in the field (Stokes, 2012). However, a few of the biodegradable mulches (e.g., boards of wood fiber, mats made of starch, or hemp fiber) were effective neither in controlling weeds nor in improving the plant growth (Stokes, 2012). An important study was conducted by Haywood (1999) where the author evaluated more than 30 natural and synthetic mulches and mats for controlling weeds in a loblolly pine (*P. taeda* L.) forest that was in its establishing phase. Several of these were effective in providing weed control lasting several seasons; regrowth of weeds was rarely observed even after the decomposition of degradable mulches; this may be due to exhaustion of weed seed bank (Haywood, 1999).

4.7 USE OF DEGRADABLE AND PAPER MULCHES

There are environmental issues caused by the use of plastic sheets as a GCS. It may not be possible to collect the used plastic mulch after the cropping season is over. Even if collected from the field, disposal of plastic sheet is a problem. The remains of plastic are indecomposable, cause hindrances in field operations (e.g., a soil applied herbicide may be restricted to completely reach the soil surface through remains of plastic mulch in the field), may disturb the growth of the following crop, and cause clogging of water channels (Steinmetz et al., 2016). Also, plastic mulch may have a negative effect on other organisms in the agricultural ecosystem. For example, the population of birds was found to decrease in a vegetable production system being put under a plastic mulch (Skorka et al., 2013). Degradable starch-based mulches have been introduced to avoid these environmental issues (Hammermeister, 2016; Kasirajan and Ngouajio, 2012). Degradable mulches may be left on the field as soon as the cropping season is over; hence, the expenditure to collect the mulch from the field after the season is over could be avoided (Goldberger et al., 2015). However, a high cost of degradable mulch may be a problem that can restrict its adoption (Goldberger et al., 2015). Use of paper mulches made from renewable stuff can also help to avoid environmental issues arising from use of plastic mulch. Compared with the uncovered soil surface, both the paper and biodegradable mulches allowed almost no weeds to grow and provided weed-free fields for higher cucumber production (Haapala et al., 2015). A comparison of plastic mulch and biodegradable plastic mulch indicated that both the mulches were effective in suppressing weeds and increasing the yield of tomato over the control (Anzalone et al., 2010).

4.8 OTHER MULCHES

There may be several other types of mulches that are used occasionally. These may include, but not limited to, compost, sawdust, peat moss, and many other by-products or waste products from some industries. For instance, in Turkey, the waste from the olive processing industry was used to control weeds and provided encouraging results (Boz et al., 2009). Boz and colleagues found that waste of olive processing could suppress several weeds in onion and faba bean (*Capsella bursa-pastoris* (L.) Medik., *Phalaris minor* Retz., *Poa annua* L., and *M. chamomilla* L.), and okra (*Amaranthus retroflexus* L., *P. oleracea*, and *Echinochloa colona* (L.) Link.). Preemergence application of olive wastes could provide weed control similar to a preemergence herbicide, while this waste had no negative effect on the crop (Boz et al., 2009). In the other investigation, Boz et al. (2010) found that olive waste was effective for controlling weeds in pea, sesame, and faba bean. Weeds such as *Melilotus officinalis* (L.) Pall., *Avena fatua* L., *A. sterilis* L., *Datura stramonium* L., *Lolium perenne* L., and *Alopecurus myosuroides* Huds. were suppressed effectively when applied with olive waste in a pot study, while no negative effect of olive waste on crops was noted (Boz et al., 2010). In the United States, hay mulch was found to suppress annual weeds along with a reduction in labor costs in tomato production (Schonbeck, 1999). In Pakistan, compost from animal manure was used to suppress weeds (and conserve soil moisture) in direct-seeded rice (personnel communication; data not published). Leaves of coconut were used as mulch for effectively suppressing weeds in ginger grown in India (Thankamani et al., 2016).

4.9 CONCLUSIONS

A variety of GCSs or mulches exist that can be employed for controlling weeds without the use of herbicides. Other than weed control, these mulches or GCSs may provide several additional benefits, soil and water conservation being the most important of these. The discussion included in this chapter provides clear evidence that several of the GCSs could provide effective weed control in vegetables mostly (this is particularly true for ones grown in greenhouses) and occasionally in fruit trees, forest tree plantations, and field crops. Environmental issues have been noted with the use of black plastic mulch; hence, the degradable mulches (e.g., those made from carbohydrates) should be preferred instead of black polyethylene. Several of the recycling materials (such as old carpets and newspapers) may also be used to suppress weeds in farming like home gardening. This will help to provide the consumers with herbicide-free and fresh vegetables.

References

Abdullah, A.S., Aziz, M.M., Siddique, K., Flower, K., 2015. Film antitranspirants increase yield in drought stressed wheat plants by maintaining high grain number. Agric. Water Manage. 159, 11–18.

Abouziena, H., Haggag, W., 2016. Weed control in clean agriculture: a review. Planta Daninha 34, 377–392.

Anzalone, A., Cirujeda, A., Aibar, J., Pardo, G., Zaragoza, C., 2010. Effect of biodegradable mulch materials on weed control in processing tomatoes. Weed Technol. 24, 369–377.

Bakht, T., Khan, I.A., 2014. Weed control in tomato (*Lycopersicon esculentum* Mill.) through mulching and herbicides. Pak. J. Bot. 46, 289–292.

Bourn, D., Prescott, J., 2002. A comparison of the nutritional value, sensory qualities, and food safety of organically and conventionally produced foods. Crit. Rev. Food Sci. Nutr. 42, 1–34.

Boyd, N., Van Acker, R., 2004. Seed germination of common weed species as affected by oxygen concentration, light, and osmotic potential. Weed Sci. 52, 589–596.

Boz, O., Ogüt, D., Kir, K., Doğan, M.N., 2009. Olive processing waste as a method of weed control for okra, faba bean, and onion. Weed Technol. 23, 569–573.

Boz, Ö., Öğüt, D., Doğan, M.N., 2010. The phytotoxicity potential of olive processing waste on selected weeds and crop plants. Phytoparasitica 38, 291–298.

Chang, D.C., Cho, J.H., Jin, Y.I., Im, J.S., Cheon, C.G., Kim, S.J., Yu, H.-S., 2016. Mulch and planting depth influence potato canopy development, underground morphology, and tuber yield. Field Crop Res. 197, 117–124.

Chaudhary, S., Iqbal, J., 2013. Weed control and nutrient promotion in zero-tillage wheat through rice straw mulch. Pak. J. Weed Sci. Res. 19, 465–474.

Chauhan, B.S., 2013. Seed germination ecology of feather lovegrass [*Eragrostis tenella* (L.) Beauv. Ex Roemer & JA Schultes]. PLoS One 8, e79398.

Chauhan, B.S., Gill, G.S., 2014. Ecologically based weed management strategies. In: Recent Advances in Weed Management. Springer, Cham, Switzerland, pp. 1–11.

Chen, H., Liu, J., Zhang, A., Chen, J., Cheng, G., Sun, B., Pi, X., Dyck, M., Si, B., Zhao, Y., 2016. Effects of straw and plastic film mulching on greenhouse gas emissions in Loess Plateau, China: a field study of 2 consecutive wheat-maize rotation cycles. Sci. Total Environ.

Crinnion, W.J., 2010. Organic foods contain higher levels of certain nutrients, lower levels of pesticides, and may provide health benefits for the consumer. Altern. Med. Rev. 15, 4–13.

Crutchfield, D.A., Wicks, G.A., Burnside, O.C., 1986. Effect of winter wheat (*Triticum aestivum*) straw mulch level on weed control. Weed Sci., 110–114.

Erenstein, O., 2003. Smallholder conservation farming in the tropics and sub-tropics: a guide to the development and dissemination of mulching with crop residues and cover crops. Agric. Ecosyst. Environ. 100, 17–37.

Farooq, M., Flower, K., Jabran, K., Wahid, A., Siddique, K.H., 2011a. Crop yield and weed management in rainfed conservation agriculture. Soil Tillage Res. 117, 172–183.

Farooq, M., Jabran, K., Cheema, Z.A., Wahid, A., Siddique, K.H., 2011b. The role of allelopathy in agricultural pest management. Pest Manage. Sci. 67, 493–506.

George, B., Brennan, P., 2002. Herbicides are more cost-effective than alternative weed control methods for increasing early growth of *Eucalyptus dunnii* and *Eucalyptus saligna*. New For. 24, 147–163.

Goldberger, J.R., Jones, R.E., Miles, C.A., Wallace, R.W., Inglis, D.A., 2015. Barriers and bridges to the adoption of biodegradable plastic mulches for US specialty crop production. Renew. Agric. Food Syst. 30, 143–153.

Haapala, T., Palonen, P., Korpela, A., Ahokas, J., 2014. Feasibility of paper mulches in crop production—a review. Agric. Food Sci. 23, 60–79.

Haapala, T., Palonen, P., Tamminen, A., Ahokas, J., 2015. Effects of different paper mulches on soil temperature and yield of cucumber (*Cucumis sativus* L.) in the temperate zone. Agric. Food Sci. 24, 52–58.

Hammermeister, A.M., 2016. Organic weed management in perennial fruits. Sci. Hortic. 208, 28–42.

Harkins, R.H., Strik, B.C., Bryla, D.R., 2014. Weed management practices for organic production of trailing blackberry. II. Accumulation and loss of biomass and nutrients. Hortscience 49, 35–43.

Hatcher, P., Melander, B., 2003. Combining physical, cultural and biological methods: prospects for integrated non-chemical weed management strategies. Weed Res. 43, 303–322.

Haywood, J.D., 1999. Durability of selected mulches, their ability to control weeds, and influence growth of loblolly pine seedlings. New For. 18, 263–276.

Haywood, J.D., 2000. Mulch and hexazinone herbicide shorten the time longleaf pine seedlings are in the grass stage and increase height growth. New For. 19, 279–290.

Hegazi, A., 2000. In: Plastic mulching for weed control and water economy in vineyards.XIVth International Symposium on Horticultural Economics 536, pp. 245–250.

Ingels, C., Lanini, T., Klonsky, K., Demoura, R., 2012. In: Effects of weed and nutrient management practices in organic pear orchards.II International Organic Fruit Symposium 1001, pp. 175–183.

Jabran, K., Farooq, M., 2013. Implications of potential allelopathic crops in agricultural systems. In: Allelopathy. Springer, Berlin Heidelberg, pp. 349–385.

Jabran, K., Ali, A., Sattar, A., Ali, Z., Yaseen, M., Iqbal, M.H.J., Munir, M.K., 2012. Cultural, mechanical and chemical weed control in wheat. Crop Environ. 3, 50–53.

Jabran, K., Mahajan, G., Sardana, V., Chauhan, B.S., 2015a. Allelopathy for weed control in agricultural systems. Crop. Prot. 72, 57–65.

Jabran, K., Ullah, E., Akbar, N., 2015b. Mulching improves crop growth, grain length, head rice and milling recovery of basmati rice grown in water-saving production systems. Int. J. Agric. Biol. 17.

Jabran, K., Ullah, E., Hussain, M., Farooq, M., Zaman, U., Yaseen, M., Chauhan, B., 2015c. Mulching improves water productivity, yield and quality of fine rice under water-saving rice production systems. J. Agron. Crop Sci. 201, 389–400.

Jabran, K., Hussain, M., Fahad, S., Farooq, M., Bajwa, A.A., Alharrby, H., Nasim, W., 2016. Economic assessment of different mulches in conventional and water-saving rice production systems. Environ. Sci. Pollut. Res. 23, 9156–9163.

Jiang, X.J., Liu, W., Wang, E., Zhou, T., Xin, P., 2017. Residual plastic mulch fragments effects on soil physical properties and water flow behavior in the Minqin oasis, northwestern China. Soil Tillage Res. 166, 100–107.

Johnson, M.S., Fennimore, S.A., 2005. Weed and crop response to colored plastic mulches in strawberry production. Hortscience 40, 1371–1375.

Kasirajan, S., Ngouajio, M., 2012. Polyethylene and biodegradable mulches for agricultural applications: a review. Agron. Sustain. Dev. 32, 501–529.

Katan, J., 2015. Soil solarization: the idea, the research and its development. Phytoparasitica 43, 1–4.

Khan, M.A., Marwat, K.B., Amin, A., Nawaz, A., Khan, R., Khan, H., Shah, H.U., 2012. Soil solarization: an organic weed-management approach in cauliflower. Commun. Soil Sci. Plant Anal. 43, 1847–1860.

Khan, I., Khan, M., Khan, M., Saeed, M., Hanif, Z., Khan, S., Ali, M., 2016. Weed management in maize (*Zea mays* L.) through different control strategies. Pak. J. Weed Sci. Res. 22, 25–35.

Kosterna, E., 2014. The effect of different types of straw mulches on weed-control in vegetables cultivation. J. Ecol. Eng. 15.

Lairon, D., 2011. Nutritional quality and safety of organic food. In: Sustainable Agriculture. vol. 2. Springer, Dordrecht, pp. 99–110.

Lament, W.J., 1993. Plastic mulches for the production of vegetable crops. HortTechnology 3, 35–39.

Liebl, R., Simmons, F.W., Wax, L.M., Stoller, E.W., 1992. Effect of rye (*Secale cereale*) mulch on weed control and soil moisture in soybean (*Glycine max*). Weed Technol., 838–846.

Lu, C., Toepel, K., Irish, R., Fenske, R.A., Barr, D.B., Bravo, R., 2006. Organic diets significantly lower children's dietary exposure to organophosphorus pesticides. Environ. Health Perspect., 260–263.

Massucati, L., Kopke, U., 2014. Effect of straw mulch residues of previous crop oats on the weed population in direct seeded faba bean in organic farming. Julius-Kühn-Archiv, 483–492.

Mo, F., Wang, J.-Y., Xiong, Y.-C., Nguluu, S.N., Li, F.-M., 2016. Ridge-furrow mulching system in semiarid Kenya: a promising solution to improve soil water availability and maize productivity. Eur. J. Agron. 80, 124–136.

Mohammed, U., Aimrun, W., Amin, M., Khalina, A., Zubairu, U., 2016. Influence of soil cover on moisture content and weed suppression under system of rice intensification (SRI). Paddy Water Environ. 14, 159–167.

Moreno, M., Moreno, A., 2008. Effect of different biodegradable and polyethylene mulches on soil properties and production in a tomato crop. Sci. Hortic. 116, 256–263.

Olson, K.D., Eidman, V.R., 1992. A farmer's choice of weed control method and the impacts of policy and risk. Rev. Agric. Econ. 14, 125–137.

Radics, L., Szné Bognár, E., 2002. In: Comparison of different mulching methods for weed control in organic green bean and tomato.XXVI International Horticultural Congress: Sustainability of Horticultural Systems in the 21st Century 638, pp. 189–196.

Rahman, M.A., Chikushi, J., Saifizzaman, M., Lauren, J.G., 2005. Rice straw mulching and nitrogen response of no-till wheat following rice in Bangladesh. Field Crop Res. 91, 71–81.

Ramakrishna, A., Tam, H.M., Wani, S.P., Long, T.D., 2006. Effect of mulch on soil temperature, moisture, weed infestation and yield of groundnut in northern Vietnam. Field Crop Res. 95, 115–125.

Razzaq, A., Cheema, Z., Jabran, K., Hussain, M., Farooq, M., Zafar, M., 2012. Reduced herbicide doses used together with allelopathic sorghum and sunflower water extracts for weed control in wheat. J. Plant Protect. Res. 52, 281–285.

Schonbeck, M.W., 1999. Weed suppression and labor costs associated with organic, plastic, and paper mulches in small-scale vegetable production. J. Sustain. Agric. 13, 13–33.

da Silva, M.G.O., de Freitas, F.C.L., Negreiros, M.Z.d., de Mesquita, H.C., de Santana, F.A.O., de Lima, M.F.P., 2013. Weed management in watermelon crop in no-tillage and conventional systems. Horticult. Brasil. 31, 494–499.

Singh, R., Ghosal, S., 2016. Effect of mulch on soil moisture, temperature, weed infestation and winter season lac yield of ber (*Ziziphus mauritiana*) under rainfed condition. Environ. Ecol. 34, 160–164.

Singh, S., Ladha, J., Gupta, R., Bhushan, L., Rao, A., Sivaprasad, B., Singh, P., 2007. Evaluation of mulching, intercropping with Sesbania and herbicide use for weed management in dry-seeded rice (*Oryza sativa* L.). Crop. Prot. 26, 518–524.

Skorka, P., Lenda, M., Moroń, D., Tryjanowski, P., 2013. New methods of crop production and farmland birds: effects of plastic mulches on species richness and abundance. J. Appl. Ecol. 50, 1387–1396.

Splawski, C.E., Regnier, E.E., Harrison, S.K., Bennett, M.A., Metzger, J.D., 2016. Weed suppression in pumpkin by mulches composed of organic municipal waste materials. Hortscience 51, 720–726.

Steinmetz, Z., Wollmann, C., Schaefer, M., Buchmann, C., David, J., Tröger, J., Muñoz, K., Frör, O., Schaumann, G.E., 2016. Plastic mulching in agriculture. Trading short-term agronomic benefits for long-term soil degradation? Sci. Total Environ. 550, 690–705.

Stokes, V., 2012. Some biodegradable mulch materials provide effective weed control during establishment of ash (*Fraxinus excelsior* L.) on farm woodland sites. Q. J. For., 257–268.

Thankamani, C., Kandiannan, K., Hamza, S., Saji, K., 2016. Effect of mulches on weed suppression and yield of ginger (*Zingiber officinale roscoe*). Sci. Hortic. 207, 125–130.

Wang, L., Gruber, S., Claupein, W., 2012. Effects of woodchip mulch and barley intercropping on weeds in lentil crops. Weed Res. 52, 161–168.

Winter, C.K., Davis, S.F., 2006. Organic foods. J. Food Sci. 71, R117–R124.

Weed Control Through Crop Plant Manipulations

Carolyn J. Lowry, Richard G. Smith
University of New Hampshire, Durham, NH, United States

5.1 INTRODUCTION

In agroecosystems, crops and weeds interact in a variety of ways, out of which, competition is the most important. Most weeds are effective competitors with crops for space, light, water, and nutrients (Zimdahl, 2004). The outcome of these competitive interactions is typically reduction in the quantity and/or quality of crop yields. Hence, competition between weeds and crops is generally considered undesirable; therefore, substantial resources have been devoted to minimizing the potential for weed-crop competition, primarily by reducing the abundance of weeds present within a crop field at the time of crop growth. Common tactics to minimize weed abundance, and therefore the potential for weed-crop competition, are to kill weeds with herbicides or the use of tillage. While herbicides can be highly effective for weed control, their use also has drawbacks, including environmental pollution (Liebman, 2001), and selection for herbicide-resistant biotypes (Vencill et al., 2012). Tillage, too, can have negative effects on soil quality and can lead to loss of soil via erosion (Kettler et al., 2000; Smith et al., 2011b). Finally, the removal of weeds by herbicides or tillage fails to address the underlying cause of weed invasion, namely, soil disturbance and overabundance of resources. Therefore, their use provides only temporary control (Smith, 2015).

An alternative approach to weed management is to manipulate the crop field environment in ways that make cropping systems less conducive to weed invasion, establishment, growth, and subsequent competition (Buhler, 2002). In this case, weed abundance is reduced because the environmental conditions and resources that support weeds are minimized or altered. This can occur via the use of crop plants that suppress weeds, are more competitive against weeds, or tolerate the presence of weeds better than others. This can also be accomplished by changing the way that cash-crop plants are planted, such that the crop population utilizes resources more efficiently over both time and space, leaving fewer resources available to support weed growth. Many of these alternative approaches also negatively impact weed fitness and fecundity. Over time, this should also have the corollary effect of reducing weed seed inputs to the soil (i.e., the soil seed bank), and therefore, the abundance of weeds that are available to infest the crop field in subsequent growing seasons (Gallandt, 2006; Nichols et al., 2015).

The benefits of weed management through the use of crop plant manipulations, particularly when multiple types of manipulations are combined, can include reduction in the need for and use of herbicides and their concomitant impacts on the environment (Liebman and Gallandt, 1997; Davis et al., 2012; Harker et al., 2016). In some cases, depending on the specific approaches that are employed, an additional benefit may be more efficient resource use by crops such that fewer fertilizer inputs are required (Anderson, 2005). Such approaches not only are necessary for cropping systems where synthetic herbicides are inaccessible (e.g., subsistence farming systems) or prohibited (e.g., certified organic cropping systems) but also may be highly beneficial to conventional cropping systems, particularly as herbicide resistance continues to be a widespread and growing problem (Heap, 2016).

The strategies for manipulating crop plants for the purpose of weed management can be placed into four broad categories. The first category is manipulations involving cover plants—that is, noncrop plants that are deliberately planted to provide soil cover and other ecosystem services, including weed suppression (Blanco-Canqui et al., 2015). These are often referred to as "cover crops," when they are grown during periods when cash crops are not growing and "living mulches" when they are utilized during periods of crop growth (Hartwig and Ammon, 2002). Hence, manipulation of cover crops aims to reduce the abundance of weeds that would potentially be available to compete with a *subsequent* cash crop, while the use of living mulches is primarily intended to minimize weed growth in the cash crop that is planted with or into the living mulch. The second category involves intercropping two or more cash crops simultaneously. Facilitative interactions between intercropped species can increase total resource utilization by the mixture, thereby increasing crop productivity and reducing resources available to stimulate weed emergence and growth. The third category involves manipulating the density or spatial orientation of cash crops so as to enhance crop competiveness and reduce the establishment and growth of weeds that are present and competing with that crop. The last category of approaches involves planting decisions based on known genetic or phenotypic variation within crop species (i.e., crop varieties or cultivars), potentially derived through breeding efforts, so as to select genotypes with superior weed suppression traits or tolerance to the presence of weeds, irrespective of the other types of manipulations (Jordan, 1993; Andrew et al., 2015).

Within these four broad categories of strategies, the agronomic literature is replete with studies that have quantified the effects of specific approaches on weed abundance and control. In many cases, thorough reviews of these strategies have already been conducted. In the sections that follow, we summarize the literature to provide selected examples illustrating the range of weed suppression provided by each category of approaches. While these approaches are often considered individually, integration of these strategies of manipulation could be expected to have additive or even synergistic (Anderson, 2005; Ryan et al., 2011a,b; Nichols et al., 2015) impacts on weed control; therefore, we highlight studies that integrate multiple approaches when these are available. Additionally, we restrict the scope of this paper primarily to the context of broadacre cropping systems in temperate agroecoregions.

5.2 MANIPULATIONS OF COVER PLANTS

Cover crops are plants typically grown during periods before cash-crop planting or after harvest, when soils would otherwise be left bare (Snapp et al., 2005). Cover crops grown during these periods can suppress weeds in a variety of ways, including by usurping soil resources and light (Blanco-Canqui et al., 2015); through the production of chemical compounds in the soil that are phytotoxic to weeds (Weston, 1996; Bhowmik and Inderjit, 2003; Farooq et al., 2011; Jabran et al., 2015); and by altering the soil and surrounding environment in ways that promote the abundance and activity of other organisms that interfere with weed growth and survival, including microorganisms (Mohler, 2001; Chee-Sanford et al., 2006; Blanco-Canqui et al., 2015) and seed predators (Gallandt et al., 2005; Shearin et al., 2008; Meiss et al., 2010; Bohan et al., 2011). In practice, these mechanisms likely often operate simultaneously to impact weed populations (Liebman and Davis, 2000; Nichols et al., 2015).

5.2.1 Weed Suppression During the Period of Cover Crop Growth

There is a large and diverse body of literature demonstrating that cover plants can suppress weeds when present in the field. The magnitude of weed suppression occurring during the period of cover-crop growth generally depends on the species of cover crop, the time it is planted, how it is terminated prior to the planting of a subsequent cash crop, and the weed species that are present (Blanco-Canqui et al., 2015). In general, previous work has demonstrated that the level of weed suppression achieved by the cover crop is highly dependent on cover-crop biomass, as larger cover crops usurp more light and soil resources from weeds (Teasdale and Mohler, 2000; Teasdale et al., 2007; Liebman and Davis, 2000). Cover crops with high vigor produce residues that are highly suppressive to weeds through a physical effect (Teasdale et al., 2007; Ryan et al., 2011a,b; Mirsky et al., 2011) and possibly higher quantities of allelochemicals (Oueslati et al., 2005; Jabran et al., 2015). Hence, the context in which the cover crop is used will ultimately determine the level of weed suppression provided by that crop.

When planted at the optimal time and under optimal growing conditions, weed suppression levels achieved by a cover crop can be nearly 100%. For example, among 12 winter cover-crop treatments (eight graminoid, legume, and radish cover-crop species grown as monocultures or bicultures) established in late summer, levels of weed suppression in the fall and into the subsequent spring, prior to cover-crop termination and cash-crop planting, were as high as 97% and 99%, respectively, depending on the species and its ability to overwinter (Hodgdon et al., 2016). In cropping systems where cover-crop establishment and growth is poor, as can occur when cover crops are planted too late or under poor planting and growing conditions, levels of weed suppression can be substantially reduced (Snapp et al., 2005; Blanco-Canqui et al., 2015). In these cases, frost seeding or overseeding of cover crops into standing crops may improve cover-crop establishment and growth (Blanco-Canqui et al., 2015).

Cover plants can also be grown during the cropping season as a smother crop or living mulch (Hartwig and Ammon, 2002). Typically, these are interseeded after cash-crop emergence, when the potential for competition with the cash crop is reduced (Hively and

Cox, 2001). In other cases, however, cash crops are planted with or into living mulches (Masiunas, 1998). While weed suppression by living mulches can be quite effective (Liebman and Dyck, 1993; Ateh and Doll, 1996), crop yields also tend to be lower in living mulch systems owing to competition between the cash and the living cover crop (Masiunas, 1998; Hartwig and Ammon, 2002; Feil and Liedgens, 2001; Hiltbrunner et al., 2007) or from changes in light quality reflected by the living mulch that trigger competition avoidance responses in the cash crop (Page et al., 2009; Warren et al., 2015).

5.2.2 Carry-Over Weed Suppression to Subsequent Cash Crops

The degree to which fall-sown cover crops continue to provide weed suppression to a subsequent cash crop (i.e., carryover weed suppression) depends on the method of termination and how the cash crop is planted. Potential mechanisms for carryover effects of preceding cover crops on weeds include reductions in weed abundance and seed rain due to competition occurring during the preceding crop (Teasdale et al., 2007; Lawley et al., 2012), production of phytotoxic chemicals that persist in the soil after the subsequent crop is planted (Bhowmik and Inderjit, 2003; Liebman and Davis, 2000), nutrient immobilization (Kuo and Sainju, 1998), the presence of smothering residues that physically inhibit weed seed germination and/or seedling emergence (Stivers-Young, 1998), or changes in soil structure that contribute to enhanced cash-crop vigor and weed suppression (Williams and Weil, 2004).

In a no-till situation, where the cover crop is terminated either mechanically (by mowing or rolling flat) or with herbicides to create a mulch layer on the soil surface, much of the subsequent weed suppression is likely due to physical suppression, although other mechanisms, such as allelopathy also likely operate (Smith et al., 2011a). One common example of this type of system involves winter rye (*Secale cereale* L.) planted in the fall and terminated in the spring with a roller crimper. A cash crop, typically soybean (*Glycine max* (L.) Merrill.), is then planted into the rolled mulch. Recent research on this system indicates that biomass levels of 8000–9000 kg ha^{-1} of rye dry matter are necessary to achieve satisfactory weed control throughout the cash-crop growing period (Smith et al., 2011a; Mirsky et al., 2013; Mischler et al., 2010).

Thus far, research is limited on the degree to which carryover weed suppression in the subsequent cash crop occurs in fall-sown cover crops that winter-kill or in cropping systems where cover crops are terminated with tillage. A recent modeling study of incorporated red clover and winter rye residues estimated that carryover weed suppression was 14% in a subsequent spring-sown cash crop (i.e., corn (*Zea mays* L.) or soybean) (Schipanski et al., 2014). While not investigated in the context of a cover crop, Anderson (2017) reported that weed abundance in alfalfa was lower when alfalfa followed soybean compared with when it followed spring wheat (*Triticum aestivum* L.) or corn, primarily due to differences in alfalfa stand establishment and vigor. Similarly, dry pea improved subsequent corn tolerance to weed competition, though the mechanism remains unknown (Anderson, 2011, 2012).

In general, the data that are available suggest that carryover weed suppression in tilled systems is substantially lower than during the period that the cover crop is alive and growing and in some cases can be nonexistent. For example, Lawley et al. (2011) observed that a fall-sown forage radish cover crop seeded at a rate of 14 kg ha^{-1} resulted in complete suppression of weeds in the spring prior to corn planting, but did not result in a carryover weed suppression benefit in the subsequent corn crop. Similarly, Hodgdon et al. (2016) observed no carryover weed suppression in a test crop planted after 11 fall-sown cover-crop treatments that had strongly suppressive effects on weeds during the previous fall and spring. In some cases, if the cover species is a legume or deep-rooted forb that scavenges nutrients that would otherwise be inaccessible to crops (Samarappuli et al., 2014), following incorporation, weed biomass levels in a subsequent crop can be higher than following some other nonlegume cover-crop treatments (Creamer et al., 1996; Hodgdon et al., 2016).

5.2.3 Mixtures or Cocktails

Typically, cover plants used as either cover crops or living mulches are sown as monocultures or simple graminoid-legume bicultures (Snapp et al., 2005); however, there is an increasing interest among growers and researchers in investigating whether there may be additional benefits to growing cover plants in more species-diverse mixtures or "cocktails" (Wortman et al., 2012; Smith et al., 2014). Two bodies of ecological theory suggest that cover plant mixtures should be more weed-suppressive than monocultures. First, diversity-productivity theory posits that productivity in plant communities is positively associated with species diversity due to species-diverse plant communities having more efficient resource use (Trenbath, 1974; Tilman, 1999). Thus, cover plant mixtures should be expected to produce more biomass compared with monocultures, and as discussed above, there is often a positive relationship observed between cover plant biomass and the level of weed suppression provided by that plant (Mohler and Teasdale, 1993; Mirsky et al., 2013).

A second complementary theory, the diversity-invasibility hypothesis, suggests that diverse cover plant mixtures would be expected to be more resistant to weed invasion than monocultures because fewer resources are left available to support weed establishment and growth (Elton, 1958; Liebman and Davis, 2000; Blanco-Canqui et al., 2015). Additionally, compared with monocultures, diverse cover plant mixtures may result in a broader spectrum of allelopathic activity toward various weed species or other modifications to the soil environment, which could further contribute to their overall weed suppressiveness (Creamer et al., 1997; Liebman and Davis, 2000).

Evidence to support the notion that weed suppression can be enhanced through the use of cover plant mixtures relative to monocultures or simple bicultures is sparse. Previous studies that have explicitly manipulated cover plant mixture diversity have found that while the mixtures often overyield relative to monocultures, weed suppression in the mixture is no higher than in the most weed-suppressive cover plant monoculture (Wortman et al., 2012; Smith et al., 2014; Finney et al., 2016). While research in this area is ongoing, the data available at present suggest that producers interested in cover plants primarily for their weed suppression benefits may be better off identifying the most weed-suppressive cover plant for their system rather than planting species-diverse mixtures.

There are a number of excellent literature reviews of cover plants, including Blanco-Canqui et al. (2015), who provide a comprehensive summary of the use of cover plants for a myriad of ecosystem services in agriculture, including weed suppression. Price and Norsworthy (2013) provide a thorough review of cover crops for weed suppression in warm-temperate regions, while Sarrantonio and Gallandt (2003) and Snapp et al. (2005) provide a review of the their use for weed control and other ecosystem services in cooler temperate regions. Ongoing research continues to identify and test "new" plant species for use as weed-suppressive cover plants (e.g., Brust et al., 2014), and nascent breeding efforts could further improve these and existing cover plant species in the area of weed management (Foley, 1999; Wayman et al., 2017).

5.3 INTERCROPPING

Intercropping is the practice of growing two or more crops or genotypes together for a part of or the whole growing season (Willey, 1990; Vandermeer, 1992) and is one mechanism by which the functional diversity of an agroecosystem can be increased (Brooker et al., 2016). For intercropping to be successful, the majority of the interactions that occur among the crop species should be beneficial and/or complementary (e.g., facilitative interactions; Brooker et al., 2008). Combining two or more crops within a mixture can sometimes increase total crop productivity because facilitative interactions among the crop species result in greater total resource utilization compared with growing the component crops as monocultures (Willey, 1990). By more efficiently utilizing resources, intercrops reduce resources available to stimulate weed emergence and growth (Liebman and Dyck, 1993).

In theory, any increase in productivity by an intercrop should result in greater weed suppression because fewer resources will remain available to weeds. Research has shown that weed suppression in intercrops can be equal to (Mohler and Liebman, 1987; Hauggaard-Nielsen et al., 2001; Deveikyte et al., 2009) or greater than (Olasantan et al., 1994; Bulson et al., 1997; Szumigalski and Van Acker, 2005; Banik et al., 2006) that attained by monocultures of the most competitive crop in the mixture. However, the effects of intercropping on crop productivity, yield, and weed suppression are often variable and dependent on the selection of appropriate crop species or genotypes (Costanzo and Barberi, 2014) and crop management (Lithourgidis et al., 2011).

Effective intercrops are often functionally diverse, meaning that the species or genotypes involved vary in their resource acquisition traits, such that complementarity between the intercropped species is maximized and competition is minimized (Tilman et al., 2001; Li et al., 2014; Brooker et al., 2015). In theory, total resource capture by the intercrop should increase as the functional diversity of the intercrop increases. Therefore, it is essential to select species or genotypes that vary in key functional traits for resource acquisition, including the spatial resource partitioning, the timing of resource uptake, and the mechanisms for resource use (Costanzo and Barberi, 2014, and references therein).

Spatial resource partitioning can occur through variations in shoot and root architecture. Utilizing crop species or genotypes that vary in shoot traits such as leaf area and angle and plant height increases the combined capture of solar radiation and reduces light penetrating the crop canopy (Keating and Carberry, 1993; Brooker et al., 2015). Complementarity in rooting depth and density can segregate root foraging of each species into distinct soil domains, thereby reducing competition for mobile resources such as N and water and increasing total soil volume exploited by the intercropped root system (Rubio et al., 2001; Zhang et al., 2014). For example, maize and squash (*Cururbita* spp.) represent ideal ideotypes for complementarity in shoot and root architecture, and intercropping squash into maize has been demonstrated to greatly reduce weed biomass without negatively affecting maize yield (Fujiyoshi et al., 2007). Maize is tall, with narrow and steeply angled leaves, allowing light to penetrate its canopy and reach the squash below. Squash is a low-growing vine that can be considered a "smother crop" because its horizontal leaves have a high leaf area (Keating and Carberry, 1993) that effectively capture excess light, thereby smothering the weed community below. Additionally, maize is shallow-rooted and forages for resources throughout the top-soil (Zhang et al., 2014), thereby reducing the availability of water and N that could stimulate weed seed germination. In contrast, squash can access deeper pools of mobile soil resources, such as N and water, which would otherwise be inaccessible to the squash (Postma and Lynch, 2012; Zhang et al., 2014).

Resource partitioning can also occur when intercropped species vary in the form or mechanisms by which soil resources are obtained. The most widely used combination of crops for intercropping is legumes combined with cereals (Yu et al., 2016). Legumes fix N, thereby increasing N available to the system (Ofori and Stern, 1987); however, most legumes are ineffective at suppressing weeds (Mohler and Liebman, 1987; Liebman and Dyck, 1993). Due to their fast growth rate and high N requirement, cereals draw down soil N concentrations, thereby stimulating legumes to increase nodulation, resulting in a greater quantity of fixed N entering the cropping system (Jensen, 1996). Fixed N may be transferred from legume nodules to the cereal crop (Ta and Faris, 1987). By alleviating N deficiency in the soil, the productivity of the cereal-legume mixture is increased, allowing greater capture of other resources such as light and water and further reducing resource availability to invading weeds. Lower weed biomass and/or density has been found in a number of different cereal/legume combinations, especially when compared with the legume crop (Hauggaard-Nielsen et al., 2001; Banik et al., 2006; Fernández-Aparicio et al., 2010; Corre-Hellou et al., 2011). One challenge with intercropping is that weed suppression may not be completely adequate to prevent yield loss. Intercropping often makes weed control difficult through other means, such as mechanical cultivation and herbicides. Often, species that are intercropped vary in their herbicide requirements and tolerance. However, sometimes, these challenges to utilizing cultivation or herbicides within intercrop systems can be overcome by manipulating the spatial orientation (through spatial arrangements) and/or timing (through relay cropping) of planting (Lithourgidis et al., 2011; Brooker et al., 2015).

Separating intercropped species into alternate strips or rows can facilitate the application of herbicides that are crop-specific and mechanization of planting, cultivation, and harvest (Exner et al., 1999). In mixed intercropping, crops are simultaneously grown together without a distinct spatial arrangement. However, with strip intercropping, space allocated to the dominant species is reduced, and the companion species gains access to a greater share of space, light, and other resources (Midmore, 1993), which may increase productivity of the less competitive species. If strips are wide enough to allow passage of cultivation equipment or prevent herbicide drift, then weed management may be separately applied to the strips of component species.

In contrast, relay intercropping is when the planting of one species in a mixture is delayed until sometime after the planting, but before harvest, of the first species (Lithourgidis et al., 2011). One benefit of relay intercropping is if the later crop is slow to establish, the first crop can act as a nurse crop by initially suppressing weeds. After the nurse crop is harvested, resources are then freed up for the second crop (Midmore, 1993). Additionally, prior to the planting of the second crop, weed management for the first crop may be optimized. Relay intercropping is also a strategy for extending the cropping season to two cash crops in regions where the climate typically limits production to a single crop (Midmore, 1993; Zhang et al., 2007). For example, in China, strip-relay intercropping is common with cotton (*Gossypium hirsutum* L.) and wheat (Zhang et al., 2007). The wheat is sown in the fall with a few rows per strip, alternated with bare soil strips where weeds can be controlled. Then, in the spring, cotton is sown into the bare soil strips.

Finally, intercropping an allelopathic crop with a nonallelopathic crop is an additional strategy to enhance functional diversity for greater weed suppression. Allelopathy is a mechanism of plant interference that occurs through the release of secondary compounds or metabolites into the environment (Weston and Duke, 2003; Jabran and Farooq, 2013). Intercropping corn with an allelopathic crop, such as squash and cowpea, was found to provide weed suppression beyond what would be predicted solely by resource preemption (Fujiyoshi et al., 2007; Jabran et al., 2015). Compared with small-seeded weeds, large-seeded cash crops, such as corn or soybean, are less likely to be susceptible to allelopathic compounds due to their lower surface area to volume ratio (Liebman and Davis, 2000). Therefore, intercropping with allelopathic crops is one strategy by which to exploit the innate size advantage between many cash crops and smaller-seeded weed species.

5.4 AGRONOMIC MANIPULATIONS OF CASH CROPS

Numerous strategies exist to manipulate planting of cash crops in order to increase their utilization of resources and therefore decrease the resources available to support weed growth. However, the optimum planting strategy depends on the morphological and physiological traits of the crop (see next section), including leaf and rooting architecture and growth rate (Gibson et al., 2003; Worthington et al., 2015). There are several ways in which crop populations can be manipulated, in terms of how they are planted, so as to take advantage of the resource acquisition traits of crops that influence weed suppression. In general, these include reducing the distance between crop rows, increasing the density of crops via seeding rate, and changing the orientation of crop rows with respect to the path of the sun.

5.4.1 Row Spacing

In crops that are planted in rows, the width between crop rows can influence both light interception by the crop canopy and soil resource uptake (Shapiro and Wortmann, 2006; Barbieri et al., 2008). In general, the weed-suppressive effects of manipulating row spacing appear to be related more to changes in the amount of light penetrating the canopy than to changes in soil resource availability (Weiner et al., 2001). In soybeans, for example, changes in row spacing can have dramatic impacts on light interception over the course of the growing season (Shibles and Weber, 1965; Puricelli et al., 2003; Steckel and Sprague, 2004). A semiquantitative review by Bradley (2006) found that in the majority of studies examined, planting soybean at a narrower row spacing reduced weed abundance compared with when soybean was planted in wider rows. Similarly, weed abundance following glyphosate application was lower in soybean with narrow rows compared with wider row spacing, and the authors attribute this effect to faster canopy closure and reduced light intensity below the canopy (Harder et al., 2007). Previous work has also demonstrated that narrower row spacing reduces weed abundance and the length of the critical period for weed control in rice (*Oryza sativa* L.) (Chauhan and Johnson, 2011), cotton (Rogers et al., 1976), snap beans (*Phaseolus vulgaris* L.) (Teasdale and Frank, 1983), and a number of other crops (Mohler, 2001).

In contrast to soybean, changing row spacing in corn has variable and often only minimal impacts on light penetration (see references in Bradley, 2006), likely due, in part, to corn having steeper leaf angles compared with soybean. Consequently, changes in row spacing in corn are not necessarily associated with changes in weed abundance (Mohler, 2001; Johnson and Hoverstad, 2002; Bradley, 2006). Interestingly, soil resource use in corn can be improved by planting in narrower spacing, and this can result in higher corn yields (Johnson and Hoverstad, 2002; Shapiro and Wortmann, 2006; Barbieri et al., 2008), which would presumably mean less nitrogen available for weeds over the longer term.

In summary, the effects of row space manipulation on weed suppression appear variable and somewhat crop species-specific (Mohler, 2001). However, in row crops, most studies that manipulate row width concomitantly also change planting density, making it difficult to separate the independent effects of these two factors (Mohler, 2001; but see Shapiro and Wortmann, 2006). In general, however, it appears that planting density (i.e., seeding rate) may be a stronger and more consistent driver of weed abundance than row spacing (Mohler, 2001).

5.4.2 Seeding Rate

Numerous studies have demonstrated that weed abundance can be reduced by planting cash crops at higher densities, primarily through changes in light competition. Mohler (2001) reviewed 91 studies involving over 25 different crops in which weed response to increasing crop density was measured. Of those, 85 of the studies (93%) found that increasing crop density was associated with decreased weed abundance. Recent work supports these findings. For example, combining doubled seeding rates of barley (*Hordeum vulgare* L.) and winter cereals with annual crop rotation reduced wild oat growth and seed production to satisfactory levels (Harker et al., 2016). Similarly, increasing the seeding rate of wheat from 400 to 600 plants m^{-2} decreased weed abundance by 30% (Kolb et al., 2012).

Given that increasing cash-crop density often reduces weed abundance, one might assume that the same would hold true for cover crops; however, there has been much less research examining the relationship between cover-crop seeding rate and weed suppression. At least in the limited number of studies that have manipulated cover-crop seeding rate, there does not appear to be a consistently strong relationship between cover-crop densities and weed suppression, perhaps owing to the fact that once sufficient ground cover has been achieved, little additional suppressive benefit is derived from additional seed. For example, while Ryan et al. (2011a) found that increasing the seeding rate of winter rye from 90 to 210 kg seed ha^{-1} reduced weed biomass in the rolled rye mulch by 31%, Hodgdon et al. (2016) found that more than doubling the seeding rate of fall-sown forage radish resulted in very little additional enhancement of weed suppression in either the fall or subsequent spring. Similarly, Smith et al.

(2015) found little improvement in weed suppression associated with a 500% increase in the seeding rate of a cover-crop mixture.

Of course, whether applied to cash crops or cover crops, the benefits of increasing crop density to weed control, and often also crop yield, must be weighed against the increased cost of seed.

5.4.3 Crop Orientation

In row crops, the spatial orientation of crop rows relative to the path of the sun can affect the competitiveness of the crop for light, and therefore, the amount of light that is available to weeds (Borger et al., 2010). Both theoretical (e.g., Palmer, 1977; Mutsaers, 1980; Schnieders et al., 1999) and empirical work (Palmer, 1977; Borger et al., 2010) suggest that crop rows that are planted at angles that are perpendicular with the path of the sun intercept more solar radiation than rows that are parallel with the sun, but that the magnitude of the effect declines with increasing latitude (Mohler, 2001). The degree to which crop row orientation will affect light interception and weed suppression is likely strongly dependent on crop leaf morphology and canopy architecture (Sarlikioti et al., 2011); therefore, the impact on weed suppression that this type of manipulation can be expected to have will be crop-specific. This was demonstrated in an elegant study by Borger et al. (2010) where they showed that planting small grain crops (wheat and barley) in rows oriented east-west reduced weed biomass by 51% and 37%, respectively, compared with the same crops grown in rows oriented north-south, and that this reduction in weed biomass was associated with increased light interception by the crop canopy. Interestingly, the researchers did not observe consistent effects of row orientation on three broadleaf crops canola (*Brassica napus* L.), field pea (*Pisum sativum*), and lupine (*Lupinus albus* L.). Similar observations of 25%–30% reductions in weed abundance due to crop row orientation have been reported in vineyards (Shrestha and Fidelibus, 2005; Alcorta et al., 2011). Recent research with wheat and barley suggests that row orientation has as much or greater of an effect on crop canopy light interception and weed fecundity as does increasing crop seeding rate (Borger et al., 2016).

Another way that light competition can be altered to reduce availability to weeds is by manipulating the spatial pattern of crop planting, thereby doing away with rows altogether. Theoretical studies suggested that a uniform pattern of planting increased the opportunity for plants to occupy space and therefore more effectively preempt light resources (Fischer and Miles, 1973). Empirical work has borne this out, indicating that sowing crops in uniform spatial patterns, rather than traditional rows, increases weed suppression in wheat, particularly when combined with higher seeding rates (Weiner et al., 2001). A recent modeling study demonstrated that uniform planting increased crop competitiveness for light by 8% compared with row planting when crops and weeds emerged simultaneously; however, effects were small relative to the timing of weed emergence (Evers and Bastiaans, 2016).

5.5 CHOICE OF CROP VARIETY OR CULTIVAR

Identification and breeding of crop varieties or cultivars that have greater competitive ability against weeds has become an emerging research priority in recent years

(Worthington and Reberg-Horton, 2013; Andrew et al., 2015). One major advantage to utilizing competitive crop varieties to suppress weeds is that compared with other integrated weed management approaches, planting weed-competitive cultivars does not require farmers to learn and implement a new practice. Farmers are often reluctant to change their farming practices (Swinton et al., 2015), and implementing some components of an integrated weed management approach (e.g., crop rotation and cover cropping) can result in additional management operations and lost opportunity costs (Snapp et al., 2005). Selecting weed-suppressive crop varieties only requires farmers to change the seeds they plant, and in fact, farmers already frequently experiment with new crop varieties. Therefore, utilizing competitive cultivars or varieties of cash crops is a relatively easy and inexpensive approach for farmers to adopt and could be a valuable strategy to suppress weeds, especially in low input or organic systems. The remainder of this section highlights some of the key traits likely important for determining variability in weed suppressiveness and tolerance to weed competition in a variety of crops.

5.5.1 Weed Suppression Versus Weed Tolerance

Crop competitive ability has two distinct components: (1) weed-suppressive ability (also called competitive effect) and (2) weed tolerance (also called competitive response) (Goldberg and Landa, 1991; Callaway, 1992). Weed-suppressive ability is a crop's capacity to interfere with the establishment, growth, and reproduction of weeds through either allelopathy or resource competition (Andrew et al., 2015). In contrast, weed tolerance is a crop's ability to not be suppressed by competing weeds or to maintain yields under weedy conditions (Callaway, 1992). The ideal variety would both suppress weeds and also maintain high yields in the presence of weed competition. However, weed-suppressive and weed-tolerant traits may (Huel and Hucl, 1996; Lemerle et al., 1996) or may not (Goldberg and Landa, 1991; Coleman et al., 2001; Didon, 2002) be correlated and are not necessarily controlled or influenced by the same mechanisms (Jordan, 1993). Therefore, in order to understand potential trade-offs between weed-suppressive and weed-tolerance capabilities, it is important to distinguish between and measure both attributes when considering competitive varieties.

5.5.2 Weed Suppression

Suppression of one plant by another, also called interference, may occur through two different mechanisms that make up the crop's "interference potential": competition and allelopathy (Worthington and Reberg-Horton, 2013). Competition includes the exploitation and depletion of necessary resources, including light, water, and soil nutrients, thereby reducing availability for competing neighbors. Allelopathy is the release of suppressive or toxic compounds into the environment that negatively affect germination or growth of susceptible plants (Weston and Duke, 2003). While competition and allelopathy are different mechanisms, they are often difficult to distinguish in the field because they can result in the same overall effect—the reduction in growth and vigor of susceptible neighbors (Bertholdsson, 2005; Worthington and Reberg-Horton, 2013). For example, an allelopathic crop cultivar with a large and extensive root system not only will have access to a greater share of soil resources but also may exude more phytochemicals (Bertholdsson, 2005). And if one cultivar released a greater quantity of allelochemicals into the environment, then weed growth would be inhibited, thereby reducing the potential for resource uptake by weeds. However, in order to select for competitive versus allelopathic traits, these mechanisms need to be understood and distinguished.

5.5.3 Weed Tolerance

In many circumstances, the ability of a crop to maintain high yields under weed competition may be desirable. Because the complete eradication of weeds is unrealistic, especially in low-input or organic cropping systems, cash crops that are better at coexisting with weed communities may result in higher yields (Ryan et al., 2009; Smith et al., 2010). While some studies have found that the traits that result in weed tolerance are the same as those important for weed suppression, including crop height and leaf area index (LAI), others believe that weed tolerance is more associated with traits related to stress resistance (Wang et al., 2010; Andrew et al., 2015). However, compared with weed suppression, weed tolerance is less likely to reduce weed populations or prevent weed infestations from increasing over time (Jordan, 1993; Murphy et al., 2008).

5.5.4 Plant Traits and Their Role in Weed Suppression and Tolerance

Competition occurs when neighboring species require the same resources at the same location and time. The ability of any one individual to outcompete another is characterized by its efficiency in acquiring, retaining, and utilizing resources (Fargione and Tilman, 2006), which is determined through a combination of morphological, physiological, and biochemical traits of that plant. Selection on competitive traits can enhance competitive ability to the degree that those traits have genetic variation and are heritable. Previous work has shown considerable variation in weed suppression among crop genotypes, and selecting more competitive cultivars can result in greater suppression of and tolerance to weeds (Worthington and Reberg-Horton, 2013). While there have been some traits that have emerged as being important for weed suppression, across studies, there has also been a considerable amount of variability and inconsistency in trait relationships to either suppression or tolerance. This variability across studies is likely due, in part, to differences in the composition and abundance of the weed communities and environmental conditions specific to each study, which underscores the importance of starting with locally adapted genotypes when selecting for more competitive cultivars.

5.5.5 Relative Growth Rate and Early Vigor

Relative growth rate (RGR), or the rate of accumulation of new dry mass per unit of existing dry mass, is a major determinant of plant competitiveness. RGR is an indirect measurement of the rate of resource acquisition, and numerous studies have found that increasing crop RGR increases weed suppression (Didon, 2002). The faster an individual accumulates biomass, the more carbon is available to increase growth of roots and shoots for greater access to light and soil nutrients, which in turn enables greater biomass accumulation. Because annual weeds typically have high RGR, especially at the seedling stage (Berger et al., 2007), enhancing crop RGR can have profound effects on resource capture and competitiveness with weeds.

5.5.6 Early Vigor: Seed Mass and Germination Rate

During crop seed germination and early seedling growth, light and soil resources are usually nonlimiting, and it is assumed that resource competition is minimal or nonexistent. However, interference and competitive responses can occur much earlier than previously believed (Ballare´ et al., 1990; Rajcan and Swanton, 2001), even before plants physically interact. This means that initial growth, occurring at the germination and seedling stages, can influence a crop plant's capacity to capture resources later, when competition for light and soil nutrients becomes more intense (Seavers and Wright, 1999). Crop varieties that have a high initial growth rate, or early vigor, are therefore able to establish earlier dominance for light and soil resources compared with varieties with lower growth rates or early vigor.

Traits related to early vigor include embryo size, seed mass, time to emergence, and seedling growth rate (Liebman and Davis, 2000; Bertholdsson, 2005 and refs therein). Seed mass is often related to the initial energy and nutrient reserves available to the crop plant during germination and establishment (Fenner, 1983). By starting with more resources, the plant can expend more energy for growth associated with acquiring more resources. In contrast to many crops, weed seeds are typically quite small (Seibert and Pearce, 1993); therefore, weed seeds often have a lower quantity of initial nutrient and energy reserves, and upon germination, weed seedlings are immediately dependent on the soil environment for their water and nutrient requirements. If soil resources are limiting, their growth is more likely to be reduced compared with the growth of larger-seeded crop seedlings (Harbur and Owen, 2006; Liebman and Davis, 2000).

Rapid seed germination and seedling emergence are also key components of early vigor and competitiveness (Harbur and Owen, 2006; Zhang et al., 2015). Faster germination results in early onset of photoautotrophic growth, enabling the early accumulation of biomass that can be utilized to access a greater share of light and belowground resources. Days to seedling emergence was correlated with increased early biomass in rice (Namuco et al., 2009) and maturation rate in sunflower (Davar et al., 2011).

5.5.7 Aboveground Traits: Improving the Plant Canopy

Shoot traits that enhance aboveground competition do so by increasing interception of light and rapid closure of the crop canopy, thereby precluding light from reaching emerging weeds. Competition for light is size asymmetrical—taller plants with larger leaves significantly reduce light reaching their shorter neighbors, while smaller plants have essentially no effect on light reaching taller neighbors (Weiner and Thomas, 1986). Interception of light is a function of a number of shoot traits, including plant height, leaf area, and leaf angle, all three of which have been found to vary considerably between crop cultivars (Andrew et al., 2015). Leaves are the photosynthetic organs of the plant; therefore, the size or area of the leaves is an important measure of a plant's capacity to capture light. Getting to and staying at the top of the canopy can be equally important as the leaf area available for light interception. Taller plants generally are more effective at suppressing weeds and tolerating weed suppression. Finally, a more horizontal or planophile leaf angle increases the quantity of light hitting the leaf surface and the efficiency of light interception (Keating and Carberry, 1993).

A number of different crops have genotypes exhibiting varying growth habits (e.g., vining vs. erect and indeterminant vs. determinant) that can greatly impact a plant's access to and interception of light. For example, Harrison and Jackson (2011) found that sweet potato cultivars with an erect shoot growth habit increased tolerance and suppression of weeds compared with a prostrate, spreading growth habit. A more erect sweet potato growth habit was characterized by shorter stems, greater branching, and a denser and taller canopy early in the growing season. In contrast, for legumes such as soybean and peanut, a more indeterminant and spreading growth habit increased weed tolerance (Fiebig et al., 1991) and weed suppression (Newcomer et al., 1986) by enabling the plant to climb over the weed canopy. Therefore, the importance of crop growth habit variation to weed suppression is likely to depend on the crop species and the abundance and species composition of the weed community.

5.5.8 Belowground Traits

Previous studies have shown that competition for belowground resources may have more impact on productivity and yield than aboveground competition (Satorre and Snaydon, 1992; Kiaer et al., 2013). In fact, species considered to be "more competitive" tend often to have a greater percent allocation to roots compared with their shoots (Aerts et al., 1991). Roots mediate plant-soil interactions; however, compared with aboveground traits, much less is known about which root traits confer a competitive advantage. This is due largely to the tedious nature of studying roots and the challenge of identifying roots to species in competition experiments. However, new advancements in studying belowground competition have led to progress on understanding the morphological and architectural traits that enhance competition for belowground resources (McNickle et al., 2008; Mommer et al., 2008; Britschgi et al., 2013; Belter and Cahill, 2015), and these insights could prove useful to future crop breeding efforts aimed at enhancing crop competitiveness with weeds.

Root RGR and architectural traits (e.g., root angles, branching pattern, number, and length of lateral and axial roots) determine where and how a crop plant can access resources in the soil, and this, in turn, affects the resources available to weeds. Crop plants with high root RGR can rapidly proliferate their root systems throughout the soil, enabling them to preempt limiting resources from potential competitors (de Kroon et al., 2003). Concentrating the bulk of the roots throughout the topsoil can preclude weed seedlings from accessing nutrients and water (Dunbabin, 2007). In contrast, production of a few roots deeper into the soil enables the crop to access deeper pools of these resources (Lynch, 2013). Root length density measures the intensity of root foraging within any given volume of soil and is directly proportional to a plant's capacity to acquire soil resources (Fargione and Tilman, 2006; Craine and Dybzinski, 2013). The more root length or surface area within a given volume of soil, the more resources will be exploited by the crop, and fewer resources will remain for invading weeds. Greater root length density is especially important if N or water are limiting (Gastal and Lemaire, 2002; Comas et al., 2013).

Because most soils are heterogeneous, crop genotypes that have high root plasticity may be able to more effectively exploit resource-rich patches. For example, in locations of the soil where water and nitrogen levels are high, increasing root uptake capacity of N and water will increase resource acquisition and RGR (Gastal and Lemaire, 2002; Comas et al., 2013). Additionally, the ability to proliferate roots into nutrient-rich patches can improve a crop's ability to preemptively usurp available nutrients from competing weeds (Robinson et al., 1999).

Root plasticity responses have recently been shown to be tied to a plant's ability to detect neighboring plants (Novoplansky, 2009). For example, some species overproliferate their roots in response to detection of a competing neighbor (Padilla et al., 2013), thereby increasing their access to soil resources and precluding the competitor from gaining access to those resources. However, if competitive cultivars are unable to distinguish between the roots of a neighboring crop plant versus those of an invading weed, then an increase in carbon allocation to roots would come at the expense of shoot growth and energy available for allocation to leaves and reproductive structures. Whether or not genetic variation for kin recognition exists within crop species remains unknown (Dudley and File, 2007).

5.5.9 Allelopathy

Allelopathic compounds are released either while the plant is still alive, via root exudation, or after the plant is dead, via decomposition of plant residues (Bhowmik and Inderjit, 2003). Typically, allelopathic compounds are produced as a plant defense mechanism to disease or herbivory (Inderjit et al., 2011). It is therefore likely that most crops have allelopathic ancestors or wild relatives and therefore the potential to produce some allelopathic compounds. Screening and selecting for allelopathy could likely increase competitive ability of most crop species. Selection efforts that focus on increasing the quantity of allelopathic compounds released via root exudates would have the greatest potential for increasing a crop's ability to suppress weeds during the growing season.

A number of crops have been found to release allelopathic root exudates, thereby increasing crop interference with weeds. For example, sorgoleone is released by roots of sorghum plants, and was found to be as effective at suppressing weeds as atrazine when applied in similar concentrations (Weston and Duke, 2003; Uddin et al., 2014). Sorghum genotypes were found to display considerable variation in the concentration of sorgoleone; therefore, selecting for increased concentration may be feasible. Other allelopathic compounds identified in crop root exudates include momilactone B in rice (Kato-Noguchi et al., 2002), hydroxamic and phenolic acids in wheat (Niemeyer and Jerez, 1997), scopoletin in oat (Fay and Duke, 1977), and gramine in barley (Hanson et al., 1981).

5.5.10 Yield-Competitiveness Tradeoff

Until recently, modern breeding efforts have largely ignored competitive traits, and breeding trials typically occur at research stations or farms with ideal weed control. Therefore, the traits that have been selected do not necessarily perform optimally in weedy conditions (Huel and Hucl, 1996). It has been suggested that continuous selection for increased yield has indirectly selected against competitive traits, implying a trade-off exists between weed competitiveness and high yields. The morphological traits involved in maintaining grain yield may differ from those that contribute to the suppression of weed growth (Coleman et al., 2001). Certain traits that confer a competitive advantage can be energetically costly to the plant, and not necessarily advantageous under weed-free conditions. For example, selecting dwarf varieties of wheat and other cereals, with reduced plant height, has resulted in reduced lodging and a greater harvest index. However, shorter varieties are often poor suppressors of weeds. This is not the case for all competitive traits, and numerous studies have found that yield-competitive trade-offs do not exist (Garrity et al., 1992; Murphy et al., 2008). Therefore, developing crop cultivars that are both high yielding and weed-suppressive is likely possible.

5.6 CONCLUSIONS

Weeds are able to invade and establish in agricultural systems when and where resources such as light, water, soil nutrients, and space are in excess of what the crop is able to utilize—essentially, weeds will invade when there is an open niche for them to do so. In monoculture systems, excess resources result from a lack of biological and functional diversity in both space and time. We have discussed four strategies to manipulate crop plants for improved weed suppression. All four strategies have in common that they strive to increase total productivity and weed suppressiveness of the crop population or crop community by increasing resource utilization.

The elimination of plant diversity in modern agricultural systems has resulted in poor resource use efficiency. We discussed two strategies by which functional diversity can be used to increase resource utilization. First, cover plants can increase functional diversity in time and can be used to usurp and sequester resources during fallow periods between cash crops. Cover plants can be extremely effective at suppressing weeds during the period of time that the cover plant is present in the field but have variable effectiveness at suppressing weeds once it is terminated. Second, we discussed intercropping as one mechanism by which to increase functional diversity in space. The effectiveness of intercropping for suppressing weeds depends on the species of crops that are grown together and crop management and environmental conditions. If the combination of crops results in facilitative interactions and complementarity in resource utilization, then overall crop productivity will likely be increased, which is likely to result in weed suppression.

Subsequently, we discussed two strategies by which the ability of the cash crop itself to suppress weeds can be enhanced. The first of these strategies was to improve the crop population's ability to capture and utilize available resources via manipulation of the crop row spacing, planting density, and row orientation. Generally speaking, decreasing the distance between crop rows, increasing planting density, and orienting crop rows so they are perpendicular to the path of the sun can increase crop light interception and result in greater weed suppression; however, this can be largely influenced by crop canopy architecture and environmental conditions. Lastly, breeding more competitive cultivars by selecting for traits that increase crop resource acquisition has the potential to greatly enhance weed suppression but has only recently become the focus of sustained research efforts.

Any one of these strategies has limited weed suppression potential. The most effective weed management will likely result from synergies arising from combining multiple of these and other management practices. For example, Ryan et al. (2011b) found that increasing the biomass of a winter rye mulch and increasing soybean density resulted in synergistic effects on weed suppression that could permit reduced soybean seeding rates and therefore potentially cost savings on seed. Similarly, intercropping maize with an allelopathic forage crop, *Desmodium uncinatum*, resulted in effective suppression of *Striga hermonthica*, a noxious parasitic weed in sub-Saharan Africa (Khan et al., 2008). Alone, the two crops have little effect on *Striga*; however, in combination, maize elicits *Striga* germination, while *Desmodium* extracts inhibit *Striga* radicle elongation, resulting in fatal germination and, over time, a drawdown of the *Striga* seed bank. Similar synergies for weed control no doubt remain to be discovered, documented, and applied on farm fields.

Acknowledgments

This work was supported by the USDA National Institute of Food and Agriculture Hatch Project 1006827, USDA NIFA AFRI Grant No. 2013-67014-21318, and USDA NIFA AFRI Grant No. 2016-67012-24677. Partial funding was provided by the New Hampshire Agricultural Experiment Station. This is scientific contribution number 2705.

References

Aerts, R., Boot, R.G.A., van der Aart, P.J.M., 1991. The relation between above- and belowground biomass allocation patterns and competitive ability. Oecologia 87, 551–559.

Alcorta, M., Fidelibus, M.W., Steenwerth, K.L., Shrestha, A., 2011. Effect of vineyard row orientation on growth and phenology of glyphosate-resistant and glyphosate-susceptible horseweed (*Conyza canadensis*). Weed Sci. 59, 55–60.

Anderson, R.L., 2017. Impact of preceding crop on alfalfa competitiveness with weeds. Renew. Agric. Food Syst. 32, 28–32.

Anderson, R.L., 2005. A multi-tactic approach to manage weed population dynamics in crop rotations. Agron. J. 97, 1579–1583.

Anderson, R.L., 2011. Synergism a rotational effect of improved growth efficiency. Adv. Agron. 112, 205–226.

Anderson, R.L., 2012. Possible causes of dry pea synergy to corn. Weed Technol. 26, 438–442.

Andrew, I.K., Storkey, J., Sparkes, D.L., 2015. A review of the potential for competitive cereal cultivars as a tool in integreated weed management. Weed Res. 55, 239–248.

Ateh, C.M., Doll, J.D., 1996. Spring-planted winter rye (*Secale cereale*) as a living mulch to control weeds in soybean (*Glycine max*). Weed Technol. 10, 347–353.

Ballare´, C.L., Scopel, A.L., Sanchez, R.A., 1990. Far-red radiation reflected from adjacent leaves: an early signal of competition in plant canopies. Science 247, 329–332.

Banik, P., Midya, A., Sarkar, B.K., Ghose, S.S., 2006. Wheat and chickpea intercropping systems in an additive series experiment: advantages and weed smothering. Eur. J. Agron. 24, 325–332.

Barbieri, P.A., Echeverría, H.E., Sainz Rozas, H.R., Andrade, F.H., 2008. Nitrogen use efficiency in no-till maize as affected by nitrogen availability and row spacing. Agron. J. 100, 1094–1100.

Belter, P.R., Cahill Jr., J.F., 2015. Disentangling root system responses to neighbors: identification of novel root behavioural strategies. AoB Plants 7, plv059.

Berger, A., McDonald, A.J., Riha, S.J., 2007. Does soil nitrogen affect early competitive traits of annual weeds in comparison with maize? Weed Res. 47, 509–516.

Bertholdsson, N.O., 2005. Early vigor and allelopathy—two useful traits for enhanced barley and wheat competitiveness against weeds. Weed Res. 45, 94–102.

Bhowmik, P.C., Inderjit, 2003. Challenges and opportunities in implementing allelopathy for natural weed management. Crop. Prot. 22, 661–671.

Blanco-Canqui, H., Shaver, T.M., Lindquist, J.L., Shapiro, C.A., Elmore, R.W., Francis, C.A., Hergert, G.W., 2015. Cover crops and ecosystem services: insights from studies in temperate soils. Agron. J. 107, 2449–2474.

Bohan, D.A., Boursault, A., Brooks, D.R., Petit, S., 2011. National-scale regulation of the weed seedbank by carabid predators. J. Appl. Ecol. 48, 888–898.

Borger, C.P.D., Hashem, A., Pathan, S., 2010. Manipulating crop row orientation to suppress weeds and increase crop yield. Weed Sci. 58, 174–178.

Borger, C.P.D., Hashem, A., Powles, S.B., 2016. Manipulating crop row orientation and crop density to suppress *Lolium rigidum*. Weed Res. 56, 22–30.

Bradley, K.W., 2006. A review of the effects of row spacing on weed management in corn and soybean. Crop Manage. https://doi.org/10.1094/CM-2006-0227-02-RV.

Britschgi, D., Stamp, P., Herrera, J.M., 2013. Root growth of neighboring maize and weeds studied with minirhizotrons. Weed Sci. 61, 319–327.

Brooker, R.W., Maestre, F.T., Callaway, R.M., Lortie, C.L., Cavieres, L.A., Kunstler, G., Liancourt, P., Tielbörger, K., Travis, J.M.J., Anthelme, F., Armas, C., Coll, L., Corcket, E., Delzon, S., Forey, E., Kikvidze, Z., Olofsson, J., Pugnaire, F., Quiroz, C.L., Saccone, P., Schiffers, K., Seifan, M., Touzard, B., Michalet, R., 2008. Facilitation in plant communities: the past, the present, and the future. J. Ecol. 96, 18–34.

Brooker, R.W., Bennett, A.E., Cong, W.F., Daniell, T.J., George, T.S., Hallett, P.D., Hawes, C., Iannetta, P.P.M., Jones, H.G., Karley, A.J., Li, L., Mckenzie, B.M., Pakeman, R.J., Paterson, E., Schöb, C., Shen, J., Squire, G., Watson, C.A., Zhang, C., Zhang, F., Zhang, J., White, P.J., 2015. Improving intercropping: a synthesis of research in agronomy, plant physiology and ecology. New Phytol. 206, 107–117.

Brooker, R.W., Karley, A.J., Newton, A.C., Pakeman, R.J., Schob, C., 2016. Facilitation and sustainable agriculture: a mechanistic approach to reconciling crop production and conservation. Funct. Ecol. 30, 98–107.

Brust, J., Claupein, W., Gergards, R., 2014. Growth and weed suppression ability of common and new cover crops in Germany. Crop. Prot. 63, 1–8.

Buhler, D.D., 2002. Challenges and opportunities for integrated weed management. Weed Sci. 50, 273–280.

Bulson, H.A.J., Snaydon, R.W., Stopres, C.E., 1997. Effect of plant density on intercropped wheat and field beans in an organic farming system. J. Agric. Sci. 128, 59–71.

Callaway, M.B., 1992. A compendium of crop varietal tolerance to weeds. Am. J. Altern. Agric. 7, 168–180.

Chauhan, B.S., Johnson, D.E., 2011. Row spacing and weed control timing affect yield of aerobic rice. Field Crop Res. 121, 226–231.

Chee-Sanford, J.C., Williams, M.M., Davis, A.S., Sims, G.K., 2006. Do microorganisms influence seed-bank dynamics? Weed Sci. 54, 575–587.

Coleman, R.D., Gill, G.S., Rebetzke, G.J., 2001. Identification of quantitative trait loci for traits conferring weed competitiveness in wheat (*Triticum aestivum* L.). Aust. J. Agric. Res. 52, 1235–1246.

Comas, L.H., Becker, S.R., Cruz, V.M.V., Byrne, P.F., Dierig, D.A., 2013. Root traits contributing to plant productivity under drought. Front. Plant Sci. 4, 1–16.

Corre-Hellou, G., Dibet, A., Hauggaard-Nielsen, H., Crozat, Y., Gooding, M., Ambus, P., Dahlmann, C., von Fragstein, P., Pristeri, A., Monti, M., Jensen, E.S., 2011. The competitive ability of pea-barley intercrops against weeds and the interactions with crop productivity and soil N availability. Food Crop Res. 122, 264–272.

Costanzo, A., Barberi, P., 2014. Functional agrobiodiversity and agroecosystem services in sustainable wheat production. a review. Agron. Sustain. Dev. 34, 327–348.

Craine, J.M., Dybzinski, R., 2013. Mechanisms of plant competition for nutrients, water and light. Funct. Ecol. 27, 833–840.

Creamer, N.G., Bennett, M.A., Stinner, B.R., Cardina, J., Regnier, E.E., 1996. Mechanisms of weed suppression in cover crop-based production systems. Hortscience 31, 410–413.

Creamer, N.G., Bennett, M.A., Stinner, B.R., 1997. Evaluation of cover crop mixtures for use in vegetable production systems. Hortscience 32, 866–870.

Davar, R., Majd, A., Darvishzadeh, R., Sarrafi, A., 2011. Mapping quantitative trait loci for seedling vigor and development in sunflower (*Helianthus annuus* L.) using recombinant inbred line population. Plant OMICS 4, 418–427.

Davis, A.S., Hill, J.D., Chase, C.A., Johanns, A.M., Liebman, M., 2012. Increasing cropping system diversity balances productivity, profitability and environmental health. PLoS One 7 (10), e47149. https://doi.org/10.1371/journal.pone.0047149.

de Kroon, H., Mommer, L., Nishiwaki, A., 2003. Root competition: toward a mechanistic understanding. In: de Kroon, H., Visser, E.J.W. (Eds.), Root Ecology. Springer, Berlin.

Deveikyte, I., Kadzuiliene, Z., Sarunaite, L., 2009. Weed suppression ability of spring cereal crops and peas in pure and mixed stands. Agron. Res. 7, 234–244.

Didon, U.M.E., 2002. Variation between barley cultivars in early response to weed competition. J. Agron. Crop Sci. 188, 176–184.

Dudley, S.A., File, A.L., 2007. Kin recognition in an annual plant. Biol. Lett. 3, 435–438.

Dunbabin, V., 2007. Simulating the role of rooting traits in crop-weed competition. Food Crop Res. 104, 44–51.

Elton, C.S., 1958. The Ecology of Invasions by Animals and Plants. Methuen, London.

Exner, D.N., Davidson, D.G., Ghaffarzadeh, M., Cruse, R.M., 1999. Yields and returns from strip intercropping on six Iowa farms. Am. J. Altern. Agric. 14, 69.

Evers, J.B., Bastiaans, L., 2016. Quantifying the effect of crop spatial arrangement on weed suppression using functional-structural plant modeling. J. Plant Res. 129, 339–351.

Fargione, J., Tilman, D., 2006. Plant species traits and capacity for resource reduction predict yield and abundance under competition in nitrogen-limited grassland. Funct. Ecol. 20, 533–540.

Farooq, M., Jabran, K., Cheema, Z.A., Wahid, A., Siddique, K.H.M., 2011. The role of allelopathy in agricultural pest management. Pest Manage. Sci. 67, 493–506.

Fay, P.K., Duke, W.B., 1977. An assessment of allelopathic potential of avena germplasm. Weed Sci. 25, 224–228.

Feil, B., Liedgens, M., 2001. Crop production in living mulches—a review. Pflanzenbauwissenshaften 5 (1), S.15–S.23.

Fenner, M., 1983. Relationships between seed weight, ash content and seed- ling growth in twenty-four species of compositae. New Phytol. 95, 697–706.

Fernández-Aparicio, M., Emeran, A.A., Rubiales, D., 2010. Intercropping with berseem clover (*Trifolium alexandrinum*) reduces infection by *Orobanche crenata* in legumes. Crop. Prot. 29, 867–871.

Fiebig, W.W., Shilling, D.G., Knauft, D.A., 1991. Peanut genotype response to interference from common cocklebur. Crop Sci. 31, 1289–1292.

Finney, D.M., White, C.M., Kaye, J.P., 2016. Biomass production and carbon/nitrogen ratio influence ecosystem services from cover crop mixtures. Agron. J. 108, 39–52.

Fischer, R.A., Miles, R.E., 1973. The role of spatial pattern in the competition between crop plants and weeds: a theoretical analysis. Math. Biosci. 18, 335–350.

Foley, M.E., 1999. Genetic approach to the development of cover crops for weed management. J. Crop. Prod. 2, 77–93.

Fujiyoshi, P.T., Gliessman, S.R., Langenheim, J.H., 2007. Factors in the suppression of weeds by squash interplanted in corn. Weed Biol. Manage. 7, 105–114.

Gallandt, E.R., 2006. How can we target the weed seedbank? Weed Sci. 54, 588–596.

Gallandt, E.R., Molloy, T., Lynch, R.P., Drummond, F.A., 2005. Effect of cover-cropping systems on invertebrate seed predation. Weed Sci. 53, 69–76.

Garrity, D.P., Movillon, M., Moody, K., 1992. Differential weed suppression ability in upland rice cultivars. Agron. J. 84, 586–591.

Gastal, F., Lemaire, G., 2002. N uptake and distribution in crops: an agronomical and ecophysiological perspective. J. Exp. Bot. 53, 789–799.

Gibson, K.D., Fischer, A.J., Foin, T.C., Hill, J.E., 2003. Crop traits related to weed suppression in water-seeded rice (*Oryza sativa* L.). Weed Sci. 51, 87–93.

Goldberg, D.E., Landa, K., 1991. Competitive effect and response hierarchies and correlated traits in early stages of competition. J. Ecol. 79, 1013–1030.

Hanson, A.D., Trayner, P.L., Dittz, K.M., Reicosky, D.A., 1981. Gramine in barley forage—effects of genotypes and environment. Crop Sci. 21, 726–730.

Harbur, M.M., Owen, M.D.K., 2006. Influence of relative time of emergence on nitrogen responses of corn and velvetleaf. Weed Sci. 54, 917–922.

Harder, D.B., Sprague, C.L., Renner, K.A., 2007. Effect of soybean row width and population on weeds, crop yield, and economic return. Weed Technol. 21, 744–752.

Harker, K.N., O'Donovan, J.T., Turkington, T.K., Blackshaw, R.E., Lupwayi, N.Z., Smith, E.G., Johnson, E.N., Pageau, D., Shirtliffe, S.J., Gulden, R.H., Rowsell, J., Hall, L.M., Willenborg, C.J., 2016. Diverse rotations and optimal cultural practices control wild oat (*Avena fatua*). Weed Sci. 64, 170–180.

Harrison, H.F., Jackson, D.M., 2011. Response of two sweet potato cultivars to weed interference. Crop. Prot. 30, 1291–1296.

Hartwig, N.L., Ammon, H.U., 2002. Cover crops and living mulches. Weed Sci. 50, 688–699.

Hauggaard-Nielsen, H., Ambus, P., Jensen, E.S., 2001. Interspecific competition, N use and interference with weeds in pea-barley intercropping. Food Crop Res. 70, 101–109.

Heap, I., 2016. The international survey of herbicide resistant weeds. Online. Internet. Thursday, December 29, 2016. Available at: www.weedscience.com.

Hiltbrunner, J., Liedgens, M., Bloch, L., Stamp, P., Streit, B., 2007. Legume cover crops as living mulches for winter wheat: components of biomass and the control of weeds. Eur. J. Agron. 26, 21–29.

Hively, W.D., Cox, W.J., 2001. Interseeding cover crops into soybean and subsequent corn yields. Agron. J. 93, 308–313.

Hodgdon, E.A., Warren, N.D., Smith, R.G., Sideman, R.G., 2016. In-season and carry-over effects of cover crops on productivity and weed suppression. Agron. J. 108, 1624–1635.

Huel, D.G., Hucl, P., 1996. Genotypic variation for competitive ability in spring wheat. Plant Breed. 115, 325–329.

Inderjit, W.D.A., Karban, R., Callaway, R.M., 2011. The ecosystem and evolutionary contexts of allelopathy. Trends Ecol. Evol. 26, 655–662.

Jabran, K., Farooq, M., 2013. Implications of potential allelopathic crops in agricultural systems. In: Cheema, Z.A., Farooq, M., Wahid, A. (Eds.), Allelopathy. Springer Berlin, Heidelberg, pp. 349–385.

Jabran, K., Mahajan, G., Sardana, V., Chauhan, B.S., 2015. Allelopathy for weed control in agricultural systems. Crop. Prot. 72, 57–65.

Jensen, E.S., 1996. Grain yield, symbiotic N2 fixation and interspecific competition for inorganic N in pea-barley intercrops. Plant Soil 182, 25–38.

Johnson, G.A., Hoverstad, T.R., 2002. Effect of row spacing and herbicide application timing on weed control and grain yield in corn (*Zea mays*). Weed Technol. 16, 548–553.

Jordan, N., 1993. Prospects for weed control through crop interference. Ecol. Appl. 3, 84–91.

Kato-Noguchi, H., Ino, T., Sata, N., 2002. Isolation and identification of a potent allelopathic substance in rice root exudates. Physiol. Plant 115, 401–405.

Keating, B.A., Carberry, P.S., 1993. Resource capture and use in intercropping: solar radiation. Food Crop Res. 34, 273–301.

Kettler, T.A., Lyon, D.J., Doran, J.W., Powers, W.L., Stroup, W.W., 2000. Soil quality assessment after weed-control tillage in a no-till wheat-fallow cropping system. Soil Sci. Soc. Am. J. 64, 339–346.

Khan, Z.R., Pickett, J.A., Hassanali, A., Hooper, A.M., Midega, C.A.O., 2008. Desmodium species and associated biochemical traits for controlling *Striga* species: present and future prospects. Weed Res. 48, 302–306.

Kiaer, L.P., Weisbach, A.N., Weiner, J., 2013. Root and shoot competition: a meta-analysis. J. Ecol. 101, 1298–1312.

Kolb, L.N., Gallandt, E.R., Mallory, E.B., 2012. Impact of spring wheat planting density, row spacing, and mechanical weed control on yield, grain protein, and economic return in Maine. Weed Sci. 60, 244–253.

Kuo, S., Sainju, U.M., 1998. Nitrogen mineralization and availability of mixed leguminous and non-leguminous cover crop residues in soil. Biol. Fertil. Soils 26, 346–353.

Lawley, Y.E., Teasdale, J.R., Weil, R.R., 2012. The mechanism for weed suppression by a forage radish cover crop. Agron. J. 104, 205–214.

Lawley, Y.E., Weil, R.R., Teasdale, J.R., 2011. Forage radish winter cover crops suppress winter annual weeds in fall and before corn planting. Agron. J. 103, 137–144.

Lemerle, D., Verbeek, B., Cousens, R.D., Coombes, N.E., 1996. The potential for selecting wheat varieties strongly competitive against weeds. Weed Res. 36, 505–513.

Li, L., Tilman, D., Lambers, H., Zhang, F.S., 2014. Plant diversity and overyielding: insights from belowground facilitation of intercropping in agriculture. New Phytol. 203, 63–69.

Liebman, M., Dyck, E., 1993. Crop rotation and intercropping strategies for weed management. Ecol. Appl. 3, 92–122.

Liebman, M., Davis, A.S., 2000. Integration of soil, crop and weed management in low-external-input farming systems. Weed Res. 40, 27–47.

Liebman, M., Gallandt, E.R., 1997. Many little hammers: ecological approaches for management of crop-weed interactions. In: Jackson, L.E. (Ed.), Ecology in Agriculture. Academic Press, San Diego, CA.

Liebman, M., 2001. Weed management: a need for ecological approaches. In: Liebman, M., Mohler, C.L., Staver, C.P. (Eds.), Ecological Management of Agricultural Weeds. Cambridge University Press, Cambridge, UK, pp. 1–39.

Lithourgidis, A.S., Dordas, C.A., Damalas, C.A., Vlachostergios, D.N., 2011. Annual intercrops: an alternative pathway for sustainable agriculture. Aust. J. Crop. Sci. 5, 396–410.

Lynch, J.P., 2013. Steep, cheap and deep: an ideotype to optimize water and N acquisition by maize root systems. Ann. Bot. 112, 347–357.

Masiunas, J.B., 1998. Production of vegetables using cover crop and living mulches—a review. J. Veg. Crop. Prod. 4, 11–31.

McNickle, G.G., Cahill, J.F., Deyholos, M.K., 2008. A PCR-based method for the identification of the roots of 10 co-occurring grassland species in mesocosm experiments. Botany 86, 485–490.

Meiss, H., Legadec, L.L., Munier-Jolain, N., Waldhardt, R., Petit, S., 2010. Weed seed predation increases with vegetation cover in perennial forage crops. Agric. Ecosyst. Environ. 138, 10–16.

Midmore, D.J., 1993. Agronomic modification of resource use and intercrop productivity. Food Crop Res. 34, 357–380.

Mirsky, S.B., Curran, W.S., Mortensen, D.M., Ryan, M.R., Shumway, D.L., 2011. Timing of cover-crop management effects on weed suppression in no-till planted soybean using a roller-crimper. Weed Sci. 59, 380–389.

Mirsky, S.B., Ryan, M.R., Teasdale, J.R., Curran, W.S., Reberg-Horton, C.S., Spargo, J.T., Wells, M.S., Keene, C.L., Moyer, J.W., 2013. Overcoming weed management challenges in cover crop-based organic rotational no-till soybean production in the eastern United States. Weed Technol. 27, 193–203.

Mischler, R.A., Curran, W.S., Duiker, S.W., Hyde, J.A., 2010. Use of a rolled-rye cover crop for weed suppression in no-till soybeans. Weed Technol. 24, 253–261.

Mohler, C.L., Liebman, M., 1987. Weed productivity and composition in sole crops and intercrops of barley and field pea. J. Appl. Ecol. 24, 685–699.

Mohler, C.L., Teasdale, J.R., 1993. Response of weed emergence to rate of *Vicia villosa* and *Secale cereale* L. residue. Weed Res. 33, 487–499.

Mohler, C.L., 2001. Weed life history: identifying vulnerabilities. In: Liebman, M., Mohler, C.L., Staver, C.P. (Eds.), Ecological Management of Agricultural Weeds. Cambridge University Press, Cambridge, UK, pp. 40–98.

Mommer, L., Wagemaker, C.A.M., de Kroon, H., Ouborg, N.J., 2008. Unravelling below-ground plant distributions: a real-time polymerase chain reaction method for quantifying species proportions in mixed root samples. Mol. Ecol. Resour. 8, 947–953.

Murphy, K.M., Dawson, J.C., Jones, S.S., 2008. Relationship among phenotypic growth traits, yield and weed suppression in spring wheat landraces and modern cultivars. Food Crop Res. 105, 107–115.

Mutsaers, H.J.W., 1980. The effect of row orientation, date and latitude on light absorption by row crops. J. Agric. Sci. 95, 381–386.

Namuco, O.S., Cairns, J.E., Johnson, D.E., 2009. Investigating early vigor in upland rice (*Oryza sativa* L.). Part I. Seedling growth and grain yield in competition with weeds. Food Crop Res. 113, 197–206.

Newcomer, D.T., Giraudo, L.J., Banks, P.A., 1986. Soybean (*Glycine max*) cultivar as a factor of weed control in no-till double-cropped production following wheat (*Triticum aestivum*). Georgia Agric. Exp. Stat. Res. Report 508, 16.

Nichols, V., Verhulst, N., Cox, R., Govaerts, B., 2015. Weed dynamics and conservation agriculture principles: a review. Field Crop Res. 183, 56–68.

Niemeyer, H.M., Jerez, J.M., 1997. Chromosomal location of genes for hydroxamic acid accumulation in *Triticum aestivum* L. (wheat) using wheat aneuploids and wheat substitution lines. Heredity 79, 10–14.

Novoplansky, A., 2009. Picking battles wisely: plant behaviour under competition. Plant Cell Environ. 32, 726–741.

Ofori, F., Stern, W.R., 1987. Cereal-legume intercropping systems. Adv. Agron. 41, 41–90.

Olasantan, F.O., Lucas, E.O., Ezumah, H.C., 1994. Effects of intercropping and fertilizer application on weed control and performance of cassava and maize. Food Crop Res. 39, 63–69.

Oueslati, O., Ben-Hammouda, M., Ghorbal, M.H., Guezzah, M., Kremer, R.J., 2005. Barley autotoxicity as influenced by varietal and seasonal variation. J. Agron. Crop Sci. 191, 249–254.

Page, E.R., Tollenaar, M., Lee, E.A., Lukens, L., Swanton, C.J., 2009. Does the shade avoidance response contribute to the critical period for weed control in maize (*Zea mays*)? Weed Res. 49, 563–571.

Padilla, F.M., Mommer, L., de Caluwe, H., Smit-Tiekstra, A.E., Wagemaker, C.A.M., Ouborg, N.J., de Kroon, H., 2013. Early root overproduction not triggered by nutrients decisive for competitive success belowground. PLoS One 8, e55805.

Palmer, J.W., 1977. Diurnal light interception and a computer model of light interception by hedgerow apple orchards. J. Appl. Ecol. 14, 601–614.

Postma, J.A., Lynch, J.P., 2012. Complementarity in root architecture for nutrient uptake in ancient maize/bean and maize/bean/squash polycultures. Ann. Bot. 110, 521–534.

Price, A.J., Norsworthy, J.K., 2013. Cover crops for weed management in southern reduced-tillage vegetable cropping systems. Weed Technol. 27, 212–217.

Puricelli, E.C., Faccini, D.E., Orioli, G.A., Sabbatini, M.R., 2003. Spurred anoda (*Anoda cristata*) competition in narrow- and wide-row soybean (*Glycine max*). Weed Technol. 17, 446–451.

Rajcan, I., Swanton, C.J., 2001. Understanding maize-weed competition: resource competition, light quality, and the whole plant. Food Crop Res. 71, 139–150.

Robinson, D., Hodge, A., Griffiths, B.S., Fitter, A.H., 1999. Plant root proliferation in nitrogen-rich patches confers competitive advantage. Proc. R. Soc. B Biol. Sci. 266, 431–435.

Rogers, N.K., Buchanan, G.A., Johnson, W.C., 1976. Influence of row spacing on weed competition with cotton. Weed Sci. 24, 410–413.

Rubio, G., Walk, T., Ge, Z., Yan, X., Liao, H., Lynch, J.P., 2001. Root gravitropism and below-ground competition among neighboring plants: a modeling approach. Ann. Bot. 88, 929–940.

Ryan, M.R., Smith, R.G., Mortensen, D.A., Teasdale, J.R., Curran, W.S., Seidel, R., Shumway, D.L., 2009. Weed and crop competition relationships differ between organic and conventional cropping systems. Weed Res. 49, 572–580.

Ryan, M.R., Mirsky, S.B., Mortensen, D.A., Teasdale, J.R., Curran, W.S., 2011b. Potential synergistic effects of cereal rye biomass and soybean planting density on weed suppression. Weed Sci. 59, 238–246.

Ryan, M.R., Curran, W.S., Grantham, A.M., Hunsberger, L.K., Mirsky, S.B., Mortensen, D.A., Nord, E.A., Wilson, D.O., 2011a. Effects of seeding rye and poultry litter on weed suppression from a rolled cereal rye cover crop. Weed Sci. 59, 438–444.

Samarappuli, D.P., Johnson, B.L., Kandel, H., Berti, M.T., 2014. Biomass yield and nitrogen content of annual energy/forage crops preceded by cover crops. Field Crop Res. 167, 31–39.

Sarlikioti, V., de Visser, P.H.B., Marcelis, L.F.M., 2011. Exploring the spatial distribution of light interception and photosynthesis of canopies by means of a functional-structural plant model. Ann. Bot. 107, 875–883.

Sarrantonio, M., Gallandt, E., 2003. The role of cover crops in North American cropping systems. J. Crop. Prod. 8, 53–74.

Satorre, E.H., Snaydon, R., 1992. A comparison of root and shoot competition between spring cereals and *Avena fatua* L. Weed Res. 32, 45–56.

Schipanski, M.E., Barbercheck, M., Douglas, M.R., Finney, D.M., Haider, K., Kaye, J.P., Kemanian, A.R., Mortensen, D.A., Ryan, M.R., Tooker, J., White, C., 2014. A framework for evaluating ecosystem services provided by cover crops in agroecosystems. Agric. Syst. 125, 12–22.

Schnieders, B.J., van der Linden, M., Lotz, L.A.P., Rabbinge, R., 1999. A model for interspecific competition in row crops. In: Schnieders, B.J. (Ed.), A Quantitative Analysis of Inter-Specific Competition in Crops With a Row Structure. Agricultural University Wageningen, Wageningen, The Netherlands, pp. 31–56.

Seavers, G.P., Wright, K.J., 1999. Crop canopy development and structure influence weed suppression. Weed Res. 39, 319–328.

Seibert, A., Pearce, R., 1993. Growth analysis of weed and crop species with reference to seed weight. Weed Sci. 41, 52–56.

Shapiro, C.A., Wortmann, C.S., 2006. Corn response to nitrogen rate, row spacing, and plant density in Eastern Nebraska. Agron. J. 98, 529–535.

Shearin, A.F., Reberg-Horton, S.C., Gallandt, E.R., 2008. Cover crop effects on the activity-density of the weed seed predator *Harpalus rufipes* (Coleoptera: Carabidae). Weed Sci. 56, 442–450.

Shibles, R.M., Weber, C.R., 1965. Leaf area, solar radiation interception and dry matter production by soybeans. Crop Sci. 5, 575–577.

Shrestha, A., Fidelibus, M., 2005. Grapevine row orientation affects light environment, growth, and development of black nightshade (*Solanum nigrum*). Weed Sci. 53, 802–812.

Smith, A.N., Reberg-Horton, C., Place, G.T., Meijer, A.D., Arellano, C., Mueller, J.P., 2011a. Rolled rye mulch for weed suppression in organic no-tillage soybeans. Weed Sci. 59, 224–231.

Smith, R.G., Mortensen, D.A., Ryan, M.R., 2010. A new hypothesis for the functional role of diversity in mediating resource pools and weed-crop competition in agroecosystems. Weed Res. 50, 37–48.

Smith, R.G., Ryan, M.R., Menalled, F.D., 2011b. Direct and indirect impacts of weed management practices on soil quality. In: Hatfield, J.L., Sauer, T.J. (Eds.), Soil Management: Building a Stable Base for Agriculture. American Society of Agronomy and Soil Science Society of America, Madison, WI, pp. 275–286.

Smith, R.G., Atwood, L.W., Warren, N.D., 2014. Increased productivity of a cover crop mixture is not associated with enhanced agroecosystem services. PLoS One 9, e97351.

Smith, R.G., Atwood, L.W., Pollnac, F.W., Warren, N.D., 2015. Cover-crop species as distinct biotic filters in weed community assembly. Weed Sci. 63, 282–295.

Smith, R.G., 2015. A succession-energy framework for reducing non-target impacts of annual crop production. Agric. Syst. 133, 14–21.

Snapp, S.S., Swinton, S.M., Labarta, R., Mutch, D., Black, J.R., Leep, R., Nyiraneza, J., O'Neil, K., 2005. Evaluating cover crops for benefits, costs and performance within cropping system niches. Agron. J. 97, 322–332.

Steckel, L.E., Sprague, C.L., 2004. Late-season common waterhemp (*Amaranthus rudis*) interference in narrow- and wide-row soybean. Weed Technol. 18, 947–952.

Stivers-Young, L., 1998. Growth, nitrogen accumulation, and weed suppression by fall cover crops following early harvest of vegetables. Hortscience 33, 60–63.

Swinton, S.M., Rector, N., Robertson, G.P., Jolejole-Foreman, C.B., Lupi, F., 2015. Farmer decisions about adopting environmentally beneficial practices. In: Hamilton, S.K., Doll, J.E., Robertson, G.P. (Eds.), The Ecology of Agricultural Landscapes: Long-Term Research on the Path to Sustainability. Oxford University Press, NY, New York, pp. 340–359.

Szumigalski, A., Van Acker, R., 2005. Weed suppression and crop production in annual intercrops. Weed Sci. 53, 813–825.

Ta, T.C., Faris, M.A., 1987. Species variation in the fixation and transfer of nitrogen from legumes to associated grasses. Plant Soil 98, 265–274.

Teasdale, J.R., Frank, J.R., 1983. Effect of row spacing on weed competition with snap beans (*Phaseolus vulgaris*). Weed Sci. 31, 81–85.

Teasdale, J.R., Mohler, C.L., 2000. The quantitative relationship between weed emergence and the physical properties of mulches. Weed Sci. 48, 385–392.

Teasdale, J.R., Brandsaeter, L.O., Calegari, A., Skora Neto, F., 2007. Cover crops and weed management. In: Upadhyaya, M.K., Blackshaw, R.E. (Eds.), Non-Chemical Weed Management: Principles, Concepts, and Technology. CAB International, Cambridge, MA, pp. 49–64.

Tilman, D., 1999. The ecological consequences of changes in biodiversity: a search for general principles. Ecology 80, 1455–1474.

Tilman, D., Reich, P.B., Knops, J., Wedin, D., Mielke, T., Lehman, C., 2001. Diversity and productivity in a long-term grassland experiment. Science 294, 843–845.

Trenbath, B.R., 1974. Biomass productivity of mixtures. Adv. Agron. 26, 177–210.

Uddin, M.R., Park, S.U., Dayan, F.E., Pyon, J.Y., 2014. Herbicidal activity of formulated sorgoleone, a natural product of sorghum root exudate. Pest Manage. Sci. 70, 252–257.

Vandermeer, J., 1992. The Ecology of Intercropping. Cambridge University Press, New York, USA.

Vencill, W.K., Nichols, R.L., Webster, T.M., Soteres, J.K., Mallory-Smith, C., Burgos, N.R., Johnson, W.G., McClelland, M.R., 2012. Herbicide resistance: toward an understanding of resistance development and the impact of herbicide-resistant crops. Weed Sci. 60, 2–30.

Wang, P., Stieglitz, T., Zhou, D.W., Cahill, J.F., 2010. Are competitive effect and response two sides of the same coin, or fundamentally different? Funct. Ecol. 24, 196–207.

Warren, N.D., Smith, R.G., Sideman, R.G., 2015. Effects of living mulch and fertilizer on the performance of broccoli in plasticulture. Hortscience 50, 218–224.

Wayman, S., Kucek, L.K., Mirsky, S.B., Ackroyd, V., Cordeau, S., Ryan, M.R., 2017. Organic and conventional farmers differ in their perspectives on cover crop use and breeding. Renew. Agric. Food Syst. 32, 376–385.

Weiner, J., Griepentrog, H.W., Kristensen, L., 2001. Suppression of weeds by spring wheat *Triticum aestivum* increases with crop density and spatial uniformity. J. Appl. Ecol. 38, 784–790.

Weiner, J., Thomas, S.C., 1986. Size variability and competition in plant monocultures. Oikos 47, 211–222.

Weston, L.A., 1996. Utilization of allelopathy for weed management in agroecosystems. Agron. J. 88, 860–866.

Weston, L.A., Duke, S.O., 2003. Weed and crop allelopathy. Crit. Rev. Plant Sci. 22, 367–389.

Willey, R.W., 1990. Resource use in intercropping systems. Agric. Water Manag. 17, 215–231.

Williams, S.M., Weil, R.R., 2004. Cover crop root channels may alleviate soil compaction effects on soybean crop. Soil Sci. Soc. Am. J. 68, 1403–1409.

Worthington, M., Reberg-Horton, C., 2013. Breeding cereal crops for enhanced weed suppression: optimizing allelopathy and competitive ability. J. Chem. Ecol. 39, 213–231.

Worthington, M., Reberg-Horton, S.C., Brown-Guedira, G., Jordan, D., Weisz, R., Murphy, J.P., 2015. Morphological traits associated with weed-suppressive ability of winter wheat against Italian ryegrass. Crop Sci. 55, 50–56.

Wortman, S.E., Francis, C.A., Lindquist, J.L., 2012. Cover crop mixtures for the western corn belt: opportunities for increased productivity and stability. Agron. J. 104, 699–705.

Yu, Y., Stomph, T.-J., Makowski, D., Zhang, L., van der Werf, W., 2016. A meta-analysis of relative crop yields in cereal/legume mixtures suggests options for management. Food Crop Res. 198, 1–11.

Zhang, L., van der Werf, W., Zhang, S., Li, B., Spiertz, J.H.J., 2007. Growth, yield and quality of wheat and cotton in relay strip intercroppping systems. Food Crop Res. 103, 178–188.

Zhang, C., Postma, J.A., York, L.M., Lynch, J.P., 2014. Root foraging elicits niche complementarity-dependent yield advantage in the ancient 'three sister' (maize/bean/squash) polyculture. Ann. Bot. 114, 1719–1733.

Zhang, J., Hu, L., Redden, B., Yan, G., 2015. Identification of fast and slow germination accessions of *Brassica napus* L. for genetic studies and breeding for early vigor. Crop Pasture Sci. 66, 481–491.

Zimdahl, R.L., 2004. Weed-Crop Competition: A Review, second ed. Blackwell Publishing, Ames, IA.

Agronomic Weed Control: A Trustworthy Approach for Sustainable Weed Management

Nicholas E. Korres
University of Arkansas, Fayetteville, AR, United States

6.1 INTRODUCTION

In recent years, non-chemical weed management approaches have gained a renewed interest due to public awareness on health issues, environmental pollution concerns, and food production cost. It has been estimated that US farmers spend annually over $3.5 billion on chemical weed control and over $2.5 billion for non-chemical weed control; the loss in food and fiber without the use of herbicides and the likely substitution of alternatives (e.g., non-chemical control methods), in values of 2003, is worth of $13.3 billion (Cahoon et al., 2017).

The rapid spread of herbicide-resistant weeds in recent years (Fig. 6.1) is mainly the result of three practices, namely, the use of monocultures, the overreliance on few single herbicide mechanisms of action, and the negligence of other weed control (Mortensen et al., 2012).

If herbicide-resistant-weed problems are addressed only with herbicides, herbicide resistance evolution will most likely prevail (Mortensen et al., 2012). Furthermore, as the number of herbicide-resistant biotypes increases, whereas the development of new modes of action declines (Strek, 2014), the need to utilize all available weed management options such as cultural, physical/mechanical, biological, and prevention tactics in conjunction with synthetic and natural originated herbicide products (Fig. 6.2) in traditional agriculture is increasing.

Another important reason that necessitates the integration of nonherbicide weed control methods into production systems is that of environmental pollution. Lutman (2018) pointed out that a number of herbicides from *Brassica napus* L. (oilseed rape) crops have the potential to leach into ground and surface waters and become transported to sources of drinking water. He also stated that where arable agriculture is a major component of water catchments, these

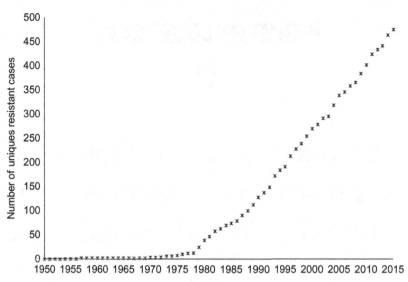

FIG. 6.1 Global increase in unique cases of herbicide-resistant weeds. *Note*: a unique case is a species × a single site of action case (Heap, 2017).

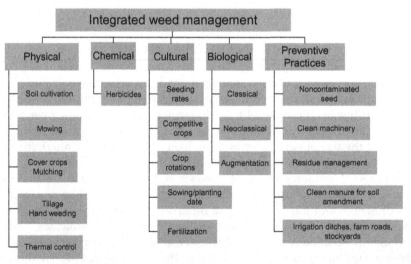

FIG. 6.2 Integrated weed management. *Based on Korres, N.E., 2005. Encyclopaedic Dictionary of Weed Science: Theory and Digest. Publishers: Lavoisier SAS; Intercept Ltd., France; UK, p. 695.*

products can greatly exceed the permissible limit in drinking water for short periods during the autumn/winter, especially when heavy rain occurs soon after herbicide application (Fig. 6.3).

Most experts agree that weed management systems in the days ahead will be more complicated (Thompson, 2012). Fortunately, the realization that weed management systems

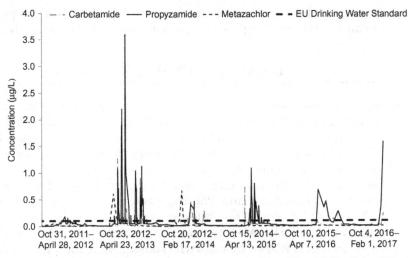

FIG. 6.3 Concentrations of herbicides detected in raw water in the River Cherwell (the United Kingdom) 2011–17, compared with the EU drinking water standard. *Adapted from Lutman, P.J.W., 2018. Sustainable weed control in oilseed rape. In: Korres, N.E., Burgos, N.R., Duke, S.O., (Eds.), Weed Control. Sustainability, Hazards and Risks in Cropping Systems Worldwide. CRC Press, ISBN:978-1498787468.*

should rely less on the panacea of herbicide products has started gaining pace in our race to revert to as many options against weeds as possible. Multitactical weed management should employ combinations from a wide range of practices that often result in synergistic or additive interactions. Farm managers and producers must make effective use of these complementary methods against herbicide resistance, even more, toward the multifaceted demand for sustainable food production (Jordan, 1996).

Agronomic weed management, in the context of this chapter, relies on the design and manipulation of cropping systems to reduce weed pressure. It can be achieved through reducing weed emergence and/or enhancing crop competitiveness providing an advantage to crop against weed competition, and it can be precisely established if the ecological characteristics of weeds and crops are known (Mohler, 1996). Cultural and physical methods (Fig. 6.2) can be used in agronomic weed control approaches. The aim of this chapter is to provide a comprehensive review for each of these methods mentioned above and to highlight the pros and cons for use in integrated weed management (IWM) programs.

6.2 AGRONOMIC WEED CONTROL METHODS

Agronomic weed control, in the context of this chapter, refers to agronomic practices that use the competitiveness of the crop to maximize its growth while diminishing the growth and subsequent competitiveness of associated weeds. Presowing or mechanical weed control methods, although very important components of the IWM systems, are excluded from this chapter. Hence, agronomic weed control refers to practices that include the use of (i) crop density, (ii) row spacing, (iii) sowing time, (iv) use of competitive cultivars, (v) crop rotations, (vi) intercropping and cover crops, and (vii) fertilization.

6.3 CROP DENSITY

Crop competitiveness against weeds can be enhanced by increasing crop density, an easily manipulating agronomic approach (Korres and Froud-Williams, 2002). At higher crop densities, crop canopy closure is accelerated resulting in reductions of the light transmitted to the soil surface and the weeds growing beneath the crop canopy (Korres and Norsworthy, 2017). This results in a decline of weed population and lower weed biomass and seed production (Korres and Norsworthy, 2017). In maize (*Zea mays* L.), increases in crop density reduce biomass of *Cyperus esculentus* L. (Ghafar and Watson, 1983) and *Abutilon theophrasti* Medik (Teasdale, 1998). Korres and Norsworthy (2015a,b) showed that increased seed rates in drill-seeded soybean (*Glycine max* (L.) Merr.) from 125,000 to 400,000 seeds/ha resulted in decreases of *Amaranthus palmeri* S. Watson biomass with subsequent reductions in seed production from 3- to 10-fold, respectively, compared with those observed in the absence of crop competition (Fig. 6.4).

Chauhan (2012) reported that increases of rice seed rate enhance crop competitiveness due to higher tiller production per unit area. Williams and Boydston (2013) recorded that increased seeding rates at maize resulted in taller and thicker crop canopy with consequent reductions in wild proso millet (*Panicum miliaceum* L.) biomass, seed production, and germinability. However, at the level of individual fields, reductions of weed growth and seed production were modest between maize populations used by growers and the higher population known to optimize the yield of certain hybrids. Responses to variation in population density demonstrate that grain yield of temperate cereal crops can be buffered against variation over a wide range of seed rates within a given growing season (Hay and Walker, 1992). Nevertheless, compensatory effects in combination with environmental conditions (Benbella and Paulsen, 1998) impose a wide range of crop yield responses to crop density manipulation. More specifically, Darwinkel (1978) investigating the effects of crop density (5, 25, 50, 100, 200, 400, and 800 plants/m^2) on growth and productivity of winter wheat (*Triticum aestivum* L.) found that optimum crop density for grain yield was 100 plants/m^2, which corresponded

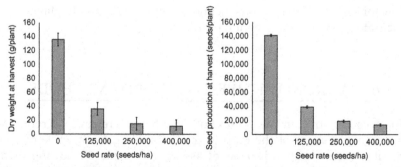

FIG. 6.4 Effects of drill-seeded soybean seed rate on *A. palmeri* dry weight (*left*) and seed production (*right*). Vertical bars represent the standard error of the mean.

to 430 ears and 19,000 grains per unit area. In addition, Smith (1980) examining a range of wheat densities observed a 5% grain yield increase between 100 and 200 plants/m^2, and he proposed as a rough guidance of optimum crop densities 125 and 150 plants/m^2 for heavy and light soils, respectively. Radford et al. (1980) testing five wheat seed rates in combination with five levels of *Avena fatua* L. infestation proposed optimum wheat density for grain yield as 53–137 and 117–252 plants/m^2 in a series of field trials over 5 years. They also highlighted the importance of increased seed rate in plots infested with wild oat and proposed a crop density of 150 plants/m^2 as a barrier for achieving satisfactory grain yield under wild oat competition. Confirming the above observation, Koscelny et al. (1990) stated that increases of seed rate in winter wheat monocultures did not affect grain yield, but under competition with *Bromus secalinus* L., crop density increases reduced yield loss due to weed presence. They proposed that increases in seed rates from 67 to 101 kg/ha reduced yield loss by 14% and 5% due to competition at two of three locations.

Increases in grain yield following increases in seed rate up to a plateau where further increases did not affect grain yield were obtained by Anderson et al. (1986). Investigating the effects of winter wheat seed rate (50%, 75%, 100%, 125%, and 150% of the normal seed rate) in combination with three row spacings on weed suppression, he observed increases in grain yield 25%–50% of the ordinary seed rate. In addition, Hashem et al. (1998), investigating the effects of seed rate of six and four winter wheat cultivars in two separate experiments on competition with *Lolium multiflorum* Lam., observed a maximum grain yield at lower crop densities (25 followed by 100 and 9 plants/m^2) in monocultures while further increases in seed rate resulted in grain yield reductions. Tollenaar et al. (1994) working with maize observed that increasing in crop density from 4 to 10 plants/m^2 was more important for weed suppression than obtaining increases in yield, in agreement with Doll et al. (1995) who stated that variation in crop density of winter cereals had a greater effect on straw production than on grain yield. Finally, Korres and Norsworthy (2015a,b) investigating the effects of three soybean seed rates (i.e., 110,000, 250,000, and 400,000 seeds/ha) in drill-seeded soybean cropping system found not significant yield increases between 250,000 and 400,000 seeds/ha.

Another important issue that needs to be addressed here is whether increasing seed rates is a suitable agronomic tool for weed control for crops other than arable crops as, for example, vegetables or perennial crops such as grapevines (*Vitis vinifera* L.) or plantations such as coffee (*Coffea arabica* L.) or pineapple (*Ananas comosus* (L.) Merr.) In vegetable crops, for example, increasing seeding rate not only can be expensive but also may enhance crop competitiveness by shading out weeds. Seeding rates in lettuce (*Lactuca sativa* L.) and spinach (*Spinacia oleracea* L.) over the past several years have increased particularly for baby leaf types (Simko et al., 2014). In the United States, baby leaf spinach, for example, can be planted at 8.64 million seeds/ha, while whole leaf fresh-cut spinach is usually planted at 3.7 million seeds/ha. In general, any crop that makes multiple units of produce on a single plant (e.g., squash, beans, and tomatoes) and most leafy greens (e.g., chard, collards, and kale) can be planted at higher than recommended rates without yield loss (Weed Ecology and Management Lab, 2017). Yield production per plant will be less at high crop densities, but the overall yield planting will be higher as crop density increases (Korres, 2005; Korres and Norsworthy, 2017). Root crops will tend to make small roots if planted too closely, although surprisingly dense plantings are possible if soil tilth and fertility are high. In plantations, such as coffee and pineapple, increased planting rates do not contribute to weed suppression during the critical period following transplanting as crop seedlings would have an initial size advantage over the weeds despite their slow growth rate compared with annual crops. Nevertheless, increasing planting densities become effective in suppressing new weed growth as the crop ages mainly due to shading (Matiello and Santinato, 2016). This could result in alterations of weed flora composition and reduction of weed biomass with possible reductions of the weed control cost (Soto-Pinto et al., 2002, Concenço et al., 2014). In addition, plantation crops that support high leaf area index values such as pineapple and banana can support high planting densities as in the case of pineapple with varying crop densities between 50,000 and 130,000 plants/ha (Zhang and Bartholomew, 1997).

6.4 ROW SPACING

Wide spacing, in monocultures, favors weed establishment and provides opportunities for weed invasions (Korres, 2005). On the contrary, reductions in row width, as much as to permit the option for interrow cultivation in soybean, resulted in increased soybean yield and increased weed suppression (Korres, 2005). In addition, Korres and Norsworthy (2015a,b) reported that interrow distance in wide-row soybean affected *A. palmeri* height, dry weight, and consequently seed production. The greater the distance from the crop, the lesser the competition effects on *A. palmeri*, which resulted in higher *A. palmeri* biomass and subsequent seed production (Fig. 6.5).

Narrowing row space and increasing seed rate enhance the competitive effects of the crop on weeds. Anderson (2000), for example, found a 60% reduction in the abundance of *Setaria italica* (L.) P. Beauv. by halving row distance and increasing seed rate by 27%, while Teasdale (1995) found a 36% reduction in weed cover by halving row spacing and doubling seed rate. Marin and Weiner (2014) compared row spacing with a spatial uniform grid pattern and found up to 75% lower biomass of the invasive weed species *Brachiaria brizantha* (A. Rich.) Stapf. Similarly, sowing of rice in narrow rows may also leave less space for weed plants (Sardana et al., 2017). Subsequently, the rice plants will be able to achieve a canopy coverage earlier than usual and shade the weeds. This shading of rice on weed plants will let the rice plants absorb more of solar radiation than weeds. Improved crop competitiveness through a

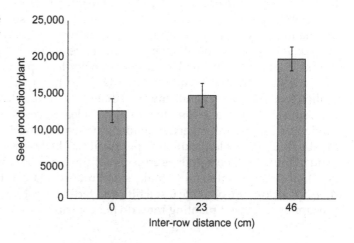

FIG. 6.5 Effects of interrow distance from the crop on *A. palmeri* seed production in wide-row soybean.

higher seed rate and/or reduced row spacing should allow for a reduction in herbicide input. Teasdale (1995) found the same effect of 25% of the recommended herbicide dose at high seed rate and narrow row spacing as of the full rate at standard seed rate and row spacing. In contrast, Dalley et al. (2004) found no significant differences in weed biomass between 38 and 76 cm row distance following application of glyphosate as a single or a sequential treatment at different timings. Similarly, Johnson et al. (1998) reported no interaction between the effect of eight herbicide treatments and row distance (51 and 76 cm) on *Setaria faberi* R. A. W. Herrm. and *Ambrosia artemisiifolia* L., and in a follow-up study, no interactions were found between row spacing and herbicide timing (Johnson and Hoverstad, 2002). Johnson and Hoverstad (2002) speculated that a higher spray interception by the crop at the narrow row spacing may offset a positive effect of narrow row spacing on crop-weed competition.

6.5 SOWING TIME

Adjustments of sowing time can influence the type and the degree of weed infestation along with the composition of the weed flora during the growing season. Vidotto et al. (2016), for example, found higher infestation levels of *Chenopodium album* L., *A. theophrasti*, and *Fallopia convolvulus* (L.) Á. Löve at early compared with conventional sowing time in maize. He also stated that early sowing time resulted at lower frequency of *Panicum dichotomiflorum* Michx., *Sorghum halepense* (L.) Pers., *Amaranthus retroflexus* L., and *Portulaca oleracea* L. Crops sown at the optimum time with adequate soil moisture and temperature, particularly in dry areas, will always be more vigorous and suppressive than those sown in less optimum conditions. If not carefully managed, early sowing might pose the risk of cool and wet soils, frost, and a greater possibility of plant disease and herbicide injury. Weed species ecology in conjunction with the potential severity of weed infestation may be used as a guide for the determination of the most appropriate time for sowing knowing that early-emerging weeds usually interfere with crops more than late-emerging weeds (Korres, 2005). Rice productivity, for example, could not be threatened due to *Echinochloa crus-galli* (L.) P. Beauv. competition, if the weed germinates after the establishment of rice crop subjected to adjustments of the rice sowing time (Gibson et al., 2002). However, when 76 versus 91 cm row spacings were compared in a potato crop, no differences were observed in either weed (*A. retroflexus*, *C. album*, and *Solanum villosum* L. Mill.) and vine biomass or tuber yield (Love et al., 1995). The authors attributed this to rapid vine elongation and a greater impact on canopy closure (Love et al., 1995). On the contrary, *C. album*, responding to the shady environment of the narrow row planting in a potato crop, increased its growth rate and biomass (Conley et al., 2001). Hence, differences in weed diversity between fields may dictate the effectiveness of using row spacing as a cultural weed management practice. It seems that delaying sowing could affect weed infestation by enhancing weed germination prior to crop sowing through the establishment of a false or stale seedbed (Buhler, 2002).

6.6 USE OF COMPETITIVE CULTIVARS

One of the main components of integrated weed management strategies for farmers is to grow more competitive crops, although selection for better crop competitiveness is a difficult task (Korres and Froud-Williams, 2002). Nevertheless, the ability of a crop to suppress weeds can be expressed in two ways either as the ability of the crop to tolerate weed competition (i.e., to maintain high yields under weedy conditions) or as the ability of the crop to suppress the growth of weeds (Callaway, 1992; Korres and Froud-Williams, 2004; Andrews et al., 2015). However, there is a confusion between cultivar tolerance to weed competition and cultivar weed-suppressive ability (Olesen et al., 2004). Furthermore, crop tolerance to weed competition varies widely over seasons and locations (Cousens and Mokhtari, 1998; Olesen et al., 2004).

For these reasons, only the cultivars that exert a suppressive ability on weeds will be discussed in this section. What makes a cultivar competitive and superior to suppress weeds? Reports have shown that a combination of characteristics, instead of a single trait, interacts for enhanced crop competitive ability. Many factors contribute to this such as early ground cover, early biomass accumulation, and rapid leaf area development of the crop in relation to weeds; ability of efficient uptake and utilization of water and nutrients; crop growth characteristics such as tallness, light penetration through the crop canopy, and shading ability; growth habit (erect vs prostrate); tillering ability; leaf width; maturity date; and potential allelopathic abilities. Additionally, the belowground characteristics such as root length density, root elongation rate, total root length, and root spatial distribution are important factors for enhancing cultivar competitive ability (Gealy et al., 2013; Fargione and Tilman, 2006; Stevanato et al., 2011). It has been reported that the greater the ability to extract water from dry soil, the greater is the competitive ability of a cultivar (Song et al., 2010). However, as stated by Rajcan and Swanton (2001), competition for water should be viewed as an outcome of the interaction between both soil-plant-atmosphere and the crop-weed systems, rather than simply as a shortage of available water. Certain crop cultivars are better competitors with weeds than others (Callaway, 1992). For example, white bean (*Phaseolus vulgaris* L.) cultivars differ in their ability to compete with weeds (Malik et al., 1993). Certain tomato cultivars (*Lycopersicon esculentum* L.) have considerable tolerance to dodder (*Cuscuta* spp.), a severe parasitic weed in many parts of the world (Goldwasser et al., 2001). Cultivars of small grain cereals with certain characteristics such as short stature, earlier maturity, better winter hardiness, or early season growth have shown differential competitive abilities when grown in mixtures compared with monocultures (Juskiw et al., 2002). In addition, as stated by Sardana et al. (2017) and Dingkuhn et al. (1991), variations in the competitive ability between rice cultivars have been documented, but those with superior morphological characteristics such as leaf area should be chosen. Nevertheless, as stated by Korres (2005), cultivar selection, as one of the most important management decisions, should also secure the production of an optimum yield. For example, work done by Korres and Froud-Williams (2002) indicated the suppressive ability of tall winter wheat cultivars on naturally occurring weed flora and their poor performance in obtaining appreciable grain yields, most probably due to low harvest index they exhibited. These authors reported that semidwarf winter wheat cultivars with high tillering ability and erect growth habit could suppress weeds better but were also capable of producing high grain yields. Similarly, Dingkuhn et al. (1999) reported that rice cultivars with high tillering ability and specific leaf area could enhance rice competitive ability.

On the other hand, maize cultivars with planophile leaves would be assumed to be more competitive than cultivars with erectophile leaves due to increased canopy light extinction, although leaf erectness can be expected to be positively correlated with maize yield. Maize cultivars that exhibit planophile leaf orientation are more prone to photosynthesis reductions at the leaves nearest to the ear, the most contributing organs of assimilates to the ear (Hammer et al., 2009). No differences detected in the amount of weed biomass accumulated through the growing season when 10 potato cultivars with contrasting growth rates and thus the amount of time between planting and canopy closure were compared (Colquhoun et al., 2009).

As stated by various authors, breeding crop cultivars with an enhanced ability to suppress weeds would be a sustainable contribution to improve weed management in many crops (Didon and Bostrom, 2003; Lemerle et al., 2001; Vollmann et al., 2010). Development of competitive crop cultivars is likely to diminish the need for mechanical weed control (Sardana et al., 2017). Developing competitive crops can be proved a cornerstone for the future delivery of sustainable weed control in crop production systems particularly in areas where climate change will most probably affect the cropping systems (Korres et al., 2016).

6.7 CROP ROTATIONS

Crop rotation is a system of growing different crops in a recurrent succession, for a specific period of time on the same land area, aiming to achieve efficient economic production with the minimum possible cost without impairing the soil fertility (Korres, 2005). Usually, the selection of the crops in the rotation program is based on market-driven considerations with the goal of maximum economic return. The sequence of the crops should also be chosen based on their requirement for minimum external inputs, nutrients, machinery, and energy to maintain soil fertility, quality, and yield. Crop variables that could affect yield include the planting and maturation dates, its competitive ability, crop growth habit, crop-specific cultural practices (i.e., the number and timing of cultivations), and fertility management (Korres, 2005).

Regarding weed management, crop rotations provide varying patterns of resource competition, soil disturbance, mechanical damage, and allelopathic interference, which result in reduced growth and reproduction of troublesome weed species. Crop sequence in the rotation (which is important because more dissimilar the crops and their management within the rotation, the fewer chances a weed species has to become dominant over several years) will dictate the use of herbicides, the tillage type, the timing of tillage relative to weed and crop emergence, and the harvest date relative to weed and crop maturity. The various weed disturbance methods in tobacco (*Nicotiana tabacum* L.) that differ to other arable crops, for example, make tobacco an effective rotation crop (Buhler et al., 1997; Liebman and Dyck, 1993).

Rotations that could restrict increases of weed soil seedbanks and prevent highly adapted weed species for becoming dominant could include the following (Korres, 2005):

- Alternations between autumn and spring germinating crops
- Alternations between annual and perennial crops
- Alternations between crops that exhibit closed and open canopies
- Alternations between cereal and broad-leaved crops to allow the use of different selective herbicides
- Choosing crops that require a variety of cultivation practices

However, crop rotations designed solely for weed control, despite the long-term benefits, may be difficult to justify based on the influence of economic and market forces on cropping management, as, for example, in maize that is usually grown in monoculture. Short-term, low profitability specialization and the lack of market opportunities are the main reasons (Lamichhane et al., 2017). Diversification of the crop rotation scheme in terms of life spans (annual, biannual, or perennial) and cropping season (winter vs summer crops) promotes the establishment of a diverse weed flora, which may allow for a reduction in herbicide input (Liebman et al., 2014; Simic et al., 2016). According to Korres (2005), the key principles of crop rotation are as follows:

(a) Tap-rooting crops should be followed by crops with a fibrous rooting pattern for proper and uniform use of nutrients and the purpose of infiltration improvement.

(b) Leguminous crops should be established after nonleguminous crops, due to the ability of the former to enrich the soil with nitrogen (e.g., crop highly demanding of nitrogen should follow a leguminous crop).

(c) Exhausting crops should be followed by less demanding crops.

(d) The marketability of the crop in rotation should be as high as possible.

(e) Crops from the same family should not be grown in succession (in this way, better disease, insect, and weed management is achieved).

(f) The farming system holds an important role in the selection of the appropriate crop to be included in the rotation program (drought-resistant crops are preferred in dry farming systems).

Besides seed rate manipulation and spatial arrangement, crop rotation is considered an important and very effective weed management strategy. Crop rotation changes weed diversity, hence preventing the development of a weed community that is perfectly tuned with the growing cycle of a particular crop (Anderson, 2007). A rotation from rice to cotton, grain sorghum (*Sorghum bicolor* (L.) Moench) or wheat reduced aquatic weed populations in flooded rice (Hill and Bayer 1990—cited in Vencill et al., 2012). Crop rotation can break the weed cycle and has been found effective in weed control in rice (Chauhan, 2012).

6.8 INTERCROPPING AND COVER CROPS

A cover crop is a crop grown between cropping seasons on arable farming systems or between orchard trees to protect the land from leaching and erosion (usually turned under for soil improvement) or to control weed establishment and growth. The inclusion of cover crops in the rotation at a time when the land might otherwise lie uncropped has been shown to be an effective method for suppressing weeds and for improving soil chemical, biological, and physical properties in various cropping systems (Alberts and Neibling, 1994; Dabney et al., 2001; Korres, 2005; Price and Norsworthy, 2013; Korres and Norsworthy, 2015a,b; Norsworthy et al., 2016). Crops such as winter vetch (*Vicia villosa* Roth) or rye (*Secale cereale* L.) can provide uniform and dense ground cover when properly managed, while crops like crown vetch (*Coronilla varia* L. (Lassen)) can provide long-term soil management. Other crops that could be used as cover crops are clovers such as crimson clover (*Trifolium incarnatum* L.), red clover (*Trifolium pratense* L.) and white clover (*Trifolium repens* L.), peas (*Pisum* spp.), bird's-foot trefoil (*Lotus corniculatus* L.), common oat (*Avena sativa* L.), ryegrasses (*Lolium* spp.), fescue (*Festuca* spp.), bluegrasses (*Poa* spp.), smooth brome (*Bromus inermis* Leyss.), timothy grass (*Phleum pratense* L.), and orchard grass (*Dactylis glomerata* L.) (Korres, 2005). Among winter crop species, winter cereals such as rye offer many benefits because they produce high biomass, are easy to establish and terminate, and provide excellent ground cover during the winter (Brown et al., 1985; Schomberg et al., 2006). Cover crops prevent weed emergence and growth through physical suppression (Akemo et al., 2000; Teasdale and Mohler, 1993; Saini et al., 2006). In cotton (*Gossypium hirsutum* L.), for example, the use of rye as a cover crop alone provided as much as 90% control of *A. retroflexus* L. due to high amounts of residue produced by the cover crop (Price et al., 2008).

Cutting cover crops (to create a living mulch) may provide adequate weed suppression through reduced light transmission and through the possible action of allelochemicals (Barnes et al., 1987; Chase et al., 1991). In oilseed rape, sowing another crop either prior to sowing the crop or between the rows is of some interest. However, the introduction of a presowing cover crop increases production costs, and its destruction prior to establishing the next crop can be challenging. According to Owen et al. (2015), adoption of the cover crops in the United States has been relatively of low interest. Nevertheless, due to the increased prevalence of glyphosate-resistant *A. palmeri* throughout the southern United States, the use of cover crops has been increasingly considered as a potential option for improved weed control (Riar et al., 2013). Cadoux et al. (2015), on the other hand, observed promising results from sowing frost-sensitive legumes between the rows of oilseed rape, as yields were not reduced by the intercrops and weed growth was less. However, a study investigating the effects of cover crops used in oilseed rape using *Fagopyrum esculentum* Moench as a cover crop showed no beneficial effects on either oilseed rape yield or weed biomass (Stumm et al., 2009).

Nevertheless, mismanagement or neglect of any cover crop can lead to additional weed problems and the potential for significant yield losses. Nevertheless, when rye was used as a cover crop for the control of glyphosate-resistant *A. palmeri*, the cover crop did not affect cotton emergence or lint yield of the crop (DeVore et al., 2012).

6.8.1 Mulches

The use of mulches as an important agronomic tool serves many purposes including enhancing plant growth by reducing soil moisture evaporation and increasing water infiltration (Watson, 1988) as well as through weed suppression (Liebl et al., 1992; Skroch et al., 1992). Usually, a living mulch consists of short-growing species established before or after the crop. It has been reported that even some weed species such as *Digitaria sanguinalis* (L.) Scop. and *P. oleracea* can be used for this purpose (Korres, 2005). Three main types of mulches can be distinguished, namely, (a) living or organic mulches, (b) mulches from plastic and other nonorganic materials, and (c) mixed mulches. Generally, the cost of mulching makes it economic only for high-value crops or perennial crops in which a mulch will remain effective for several years, while the main advantages of mulching include early cropping, reduced weed pressure, and reduced evaporation from the soil surface.

Intercropping is common in Africa, South America, and some parts of Asia. The underlying concept of intercropping is to increase yields by making better use of light, water, and nutrients (Vandermeer, 1989). A potential benefit of intercropping is a better suppression of weeds, while, on the other hand, intercropping may render the use of physical and chemical weed control methods impossible or very difficult. As maize is often grown at wide row spacing, it is very suitable for intercropping, and many studies have examined maize intercropping, in particular, with legume crops. The results reported by Jamshidia et al. (2013) illustrate the potential benefits of intercropping. In their study, maize was intercropped with cowpea (*Vigna unguiculata* (L.) Walp.), and the inclusion of cowpea reduced weed biomass by ca. 40% compared with maize alone. Under weed-free conditions, intercropping with cowpea had a negligible effect on maize yield (by up to 4.2%), but, under weed-infested conditions, maize yield was increased by up to 32%.

In Africa, one of the main drivers for intercropping is the management of *Striga asiatica* (L.) Kuntze and *Striga hermonthica* (Delile) Benth. and of the stem borer (Woomer et al., 2008). Several legumes have been studied, but the most successful in terms of both *Striga* control and net profit is fodder legumes belonging to the genus *Desmodium* that are used in a so-called "push-pull" strategy (Khan et al., 2006; Midega et al., 2014). This is a chemical-based approach composed of the exudation of compounds preventing the attachment of *Striga* to the maize roots and subsequent emergence (Khan et al., 2016).

6.9 NITROGEN FERTILIZATION

Weed-crop competition for nutrients, especially for nitrogen, is one of the most important problems since the availability of nitrogen is often the limiting factor in plant growth especially in soils with low supplementary ability. The factors that account for nitrogen variation in crop systems are soil type, soil organic content and availability of water, seasonal precipitation, date of sowing, choice of variety, rate and application of nitrogen fertilization, and weed control. Nitrogen management practices can affect the outcome of competition with respect to the weed population and its competitive ability relative to the crop (Evans et al., 2003; Wortman et al., 2011). It has been reported that broadcast nitrogen application stimulates the growth of *Bromus tectorum* L. more than does deep band placement in a fallow wheat system (Ball et al., 1996). Research on the effects of competition for nitrogen related to crop response has shown that tall cultivars of wheat subject to relatively high fertilizer rates could compete more efficiently with *A. fatua* due to competition for light (Gonzalez-Ponce and Santin, 1999). Reductions of nitrogen, phosphorus, and potassium in a range of cereal crops due to *Lolium rigidum* Gaudin, *Veronica hederifolia* L., and *A. fatua* competition have also been reported (Lemerle et al., 1995; Angonin et al., 1996). Another factor that determines the outcome of competition for nutrients is the phenotypic plasticity of weeds (i.e., species-dependent) that enables them to utilize high nutrient levels by means of luxuriant growth (Blackshaw et al., 2003). It has been shown that the competitive effects of weeds such as *Bromus sterilis* L., *Cyperus rotundus* L., *Alopecurus myosuroides* Huds., *Galium aparine* L., *A. fatua*, and *A. retroflexus* increased at greater amounts of nitrogen application under various cereal crops (Lintell-Smith et al., 1992; Okafor and De Datta, 1976; Grundy and Froud-Williams, 1991; Wilson, 1999; Wright, 1993; Teyker et al., 1991). In contrast, some other weed species show an ability to grow better than others in soil with low levels of nutrients. Weeds such as *Matricaria inodora* Lam. and *Polygonum aviculare* L. exerted the strongest competition in the absence of nitrogen fertilizer but were progressively suppressed by the crop as the nitrogen application rate increased (Scott and Wilcockson, 1976; Iqbal and Wright, 1997).

Additionally, it has been shown that *Stellaria media* (L.) Vill., *C. album*, *Cirsium arvense* (L.) Scop., and *Elytrigia repens* (L.) Nevski can accumulate less nitrogen than cereals (Jornsgard et al., 1996; Stupnicka-Rodzynkiewicz et al., 1999). Appropriate timing of nitrogen application has been proposed in integrated systems as a means of unbalancing nutrient competition between crop and weeds, to benefit the former (Liebman and Davis, 2000; Jabran and Chauhan, 2015; Korres et al., 2017) particularly for these weed species that thrive at high nitrogen environments such as *A. palmeri* (Korres et al., 2017). This approach necessitates knowledge of the weed flora that is likely to develop in a field (Korres, 2005). Nevertheless, Maqbool et al. (2016) found no significant differences in weed biomass or maize yield at various fertilizer placement although the below seed treatment tended to produce lower weed biomass and higher yields. The rate of nutrient release by organic fertilizers and amendments in organic farming systems depends on the C/N ratio of the source, soil properties, climatic conditions, and method of incorporation (Korres, 2005). Faster release rates favor the weeds, which are usually capable of taking up and utilizing nutrients at earlier growth stages more quickly and efficiently than the crop, although most probably, this effect turns into a competitive advantage when weed density is high. In contrast, slower release rates may accelerate the occurrence of late-season weed emergence that contributes to seedbank replenishment and consequently to higher weed seedling recruitment in subsequent years (Korres, 2005).

6.10 CONCLUSION

Although the agronomic practices described above are not without costs and risks (some of them are shown in Table 6.1) (Counce and Burgos, 2006), they have been proved useful tools particularly in situations where herbicide application options are restricted.

TABLE 6.1 Agronomic Practices for Incorporation in Weed Management Systems and Associated Costs and Risks

Agronomic Practice	Costs	Risks
Fertilization optimum nitrogen	Extra N cost	Lower yield for inadequate N, lodging for excessive N
Cultivar selection	Extra seed costs	Lower yields from some more competitive cultivars
Reduced row spacing	Cost for new planting equipment	Increased lodging
Increased seed rate	Increased cost for seed purchase	Increased lodging
Cover crop	Direct (e.g., cover crop seed, labor involved in planting, and "terminating" the crop) and indirect costs (cover crop in expense of cash crops)	Return of investment difficult to be anticipated against the immediate cost of running a farm

Based on Counce, P.A., Burgos, N.R., 2006. Agronomic practices in addition to herbicides for improving weed control in dry-seeded flooded rice. J. Sustain. Agric. 28 (3), 145–156.

To complement the use of herbicides, these strategies aim to enhance the competitive advantage of the crop or to act either synergistically or additively resulting in effective weed control. Farmers should broaden and diversify their weed control options by incorporating suitable agronomic techniques into their current production systems, ultimately reducing the risk for new cases of herbicide resistance evolution.

References

Akemo, M.C., Regnier, E.E., Bennett, M.A., 2000. Weed suppression in spring-sown rye (*Secale cereale*) pea (*Pisum sativum*) cover crop mixes. Weed Technol. 14, 545–549.

Alberts, E.E., Neibling, W.H., 1994. Influence of crop residues on water erosion. In: Unger, P.W. (Ed.), Managing Agricultural Residues. Lewis Publishing, Ann Arbor, MI, pp. 19–39.

Anderson, R.L., 2000. Cultural systems to aid weed management in semiarid corn (*Zea mays*). Weed Technol. 14, 630–634.

Anderson, R.L., 2007. Managing weeds with a dualistic approach of prevention and control. Agron. Sustain. Dev. 27, 13–18.

Anderson, A., Radosevich, S.R., Roush, M.L., 1986. In: Influence of crop density and spacing on weed competition and grain yield in wheat and barley. Proceedings of European Weed Research Society, pp. 121–128.

Andrews, I.K.S., Storkey, J., Sparkes, D.L., 2015. A review of the potential for competitive cereal cultivars as a tool in integrated weed management. Weed Res. 55, 239–248.

Angonin, C., Caussanel, J.P., Maynard, J.M., 1996. Competition between winter wheat and *Veronica hederifolia*: influence of weed density and the amount and timing of nitrogen application. Weed Res. 36, 175–187.

Ball, D.A., Wysocki, D.J., Chastain, T.G., 1996. Nitrogen application timing effects on downy brome (*Bromus tectorum*) and winter wheat (*Triticum aestivum*) growth and yield. Weed Technol. 10, 305–310.

Barnes, J.P., Putman, A.R., Burke, B.A., Aasen, A.J., 1987. Isolation and characterization of allelochemicals in rye herbage. Phytochemistry 26, 1385–1390.

Benbella, M., Paulsen, G.M., 1998. Efficacy of treatments for delaying senescence of wheat leaves: II. senescence and grain yield under field conditions. Agron. J. 90, 332–338.

Blackshaw, R.E., Brandt, R.N., Jazen, H.H., Entz, T., Grant, C., Derksen, A., 2003. Differential response of weed species to added nitrogen. Weed Sci. 52, 532–539.

Brown, S.M., Whitwell, T., Touchton, J.T., Burmester, C.H., 1985. Conservation tillage systems for cotton production (USA). Soil Sci. Soc. Am. J. 49, 1256–1260.

Buhler, D.D., 2002. Challenges and opportunities for integrated weed management. Weed Sci. 50, 273–280.

Buhler, D.D., Hartzler, R.G., Forcella, F., 1997. Implications of weed seedbank dynamics to weed management. Weed Sci. 45, 329–336.

Cadoux, S., Sauzet, G., Valantin-Morison, M., Pontet, C., Champolivier, L., Robert, C., Lieven, J., Flenet, F., Mangenot, O., Fauvin, P., Lande, N., 2015. Intercropping frost-sensitive legume crops with winter oilseed rape reduces weed competition, insect damage, and improves nitrogen use efficiency. OCL 22 (D302), 11.

Cahoon, C.W., Flessner, M.L., Shultz, B., Curran, W.S., Chandran, R., Van-Gessel, M., Lingenfelter, D., Johnson, Q., Vollmer, K., Hines, T., 2017. Weed control in field crops introduction to weeds and weed management. In: Pest Management Guide, Field Crops 2017. Virginia Cooperative Extension, Virginia, USA, pp. 5(1)–5(355). Pub. 456-016.

Callaway, M.B., 1992. A compendium of crop varietal tolerance to weeds. Am. J. Altern. Agric. 7, 169–180.

Chase, W.R., Nair, M.G., Putman, A.R., 1991. 2,2′-Oxo-1,1′-azobenzene: selective toxicity of rye (*Secale cereale* L.) allelochemicals to weed and crop species: II. J. Chem. Ecol. 19, 9–19.

Chauhan, B.S., 2012. Weed ecology and weed management strategies for dry-seeded rice in Asia. Weed Technol. 26, 1–13.

Colquhoun, J.B., Konieczka, C.M., Rittmeyer, R.A., 2009. Ability of potato cultivars to tolerate and suppress weeds. Weed Technol. 23, 287–291.

Concenço, G., Motta, I.S., Correia, I.V.T., Santos, S.A., Mariani, A., Marques, R.F., Palharini, W.G., Alves, M.E.S., 2014. Infestation of weed species in monocrop coffee or intercropped with banana, under agroecological system. Planta daninha. 32, 665–674.

Conley, S.P., Binning, L.I., Connell, T.R., 2001. Effect of cultivar, row spacing, and weed management on weed biomass, potato yield, and net crop value. Am. J. Potato Res. 78, 31–37.

Counce, P.A., Burgos, N.R., 2006. Agronomic practices in addition to herbicides for improving weed control in dry-seeded flooded rice. J. Sustain. Agric. 28 (3), 145–156.

Cousens, R.D., Mokhtari, S., 1998. Seasonal and site variability in the tolerance of wheat cultivars to interference from *Lolium rigidum*. Weed Res. 38, 301–307.

Dabney, S.M., Delgado, J.A., Reeves, D.W., 2001. Using winter cover crops to improve soil and water quality. Commun. Soil Sci. Plant Anal. 32, 1221–1250.

Dalley, C.D., Kells, J.J., Renner, K.A., 2004. Effect of glyphosate application timing and row spacing on weed growth in corn (*Zea mays*) and soybean (*Glycine max*). Weed Technol. 180, 177–182.

Darwinkel, A., 1978. Patterns of tillering and grain production of winter wheat at a wide range of plant densities. Neth. J. Agric. Sci. 26, 383–398.

DeVore, J.D., Norsworthy, J.K., Brye, K.R., 2012. Influence of deep tillage and cover crop on glyphosate-resistant Palmer amaranth (*Amaranthus palmeri* L. Wats) emergence in cotton. Weed Technol. 26, 832–838.

Didon, U.M.E., Bostrom, U., 2003. Growth and development of six barley (*Hordeum vulgare* ssp. *vulgare* L.) cultivars in response to a model weed (*Sinapis alba* L.). J. Agron. Crop Sci. 189, 409–417.

Dingkuhn, M., Penning de Vries, F.W.T., De Datta, S.K., Van Laar, H.H., 1991. Concept for a new plant type for direct seeded flooded tropical rice. In: Direct Seeded Flooded Rice in the Tropics. Selected papers from the Int. Rice Research Conf. August 1990, Seoul, Korea. Int. Rice Res. Inst. 1991, Los B&OS, pp. 7–38.

Dingkuhn, M., Johnson, D.E., Sow, A., Audebert, A.Y., 1999. Relationships between upland rice canopy characteristics and weed competitiveness. Field Crop Res. 61, 79–95.

Doll, H., Holm, U., Sogaard, B., 1995. Effects of crop density on competition by wheat and barley with *Agrostemma githago* and other weeds. Weed Res. 35, 391–396.

Evans, S.P., Knezevic, S.Z., Lindquist, J.L., Shapiro, C.A., 2003. Influence of nitrogen and duration of weed interference on corn growth and development. Weed Sci. 51, 546–556.

Fargione, J., Tilman, D., 2006. Plant species traits and capacity for resource reduction predict yield and abundance under competition in nitrogen-limited grassland. Funct. Ecol. 20, 533–540.

Gealy, D.R., Moldenhauer, K.A., Duke, S., 2013. Root distribution and potential interactions between allelopathic rice, sprangletop (*Leptochloa* spp.), and barnyardgrass (*Echinochloa crus-galli*) based on ^{13}C isotope discrimination analysis. J. Chem. Ecol. 39, 186–203.

Ghafar, Z., Watson, A.K., 1983. Effect of corn (*Zea mays*) population on the growth of yellow nutsedge (*Cyperus esculentus*). Weed Sci. 31, 588–592.

Gibson, K., Fischer, A., Foin, T., Hill, J., 2002. Implications of delayed *Echinochloa* spp. germination and duration of competition for integrated weed management in water-seeded rice. Weed Res. 42, 351–358.

Goldwasser, Y., Lanini, W.T., Wrobel, R.L., 2001. Tolerance of tomato varieties to lespedeza dodder. Weed Sci. 49, 520–523.

Gonzalez-Ponce, R., Santin, I., 1999. In: Effect of N fertilisation on competition of wheat with wild oats. Proceedings of 11 European Weed Research Society Symposium, p. 56.

Grundy, A.C., Froud-Williams, R.J., 1991. In: The effects of herbicide and fertiliser rate on weed productivity in spring wheat.Brighton Crop Protection Conference-Weeds, vol. 1, pp. 411–418.

Hammer, G.L., Dong, Z., McLean, G., Doherty, A., Messina, C., Schussler, J., Zinselmeier, C., Paszkiewicz, S., Cooper, M., 2009. Can changes in canopy and/or root system architecture explain historical maize yield trends in the U.S. Corn Belt? Crop Sci. 49, 299–312.

Hashem, A., Radosevich, S.R., Roush, M.L., 1998. Effect of proximity factors on competition between winter wheat (*Triticum aestivum*) and Italian ryegrass (*Lolium multiflorum*). Weed Sci. 46, 181–190.

Hay, J.L., Walker, A.J., 1992. An Introduction to the Physiology of Crop Yield. Longman Scientific and Technical & John Wiley and sons Publications, Oxford, UK, pp. 163–168.

Heap, I., 2017. International Survey of Herbicide Resistant Weeds. Available at: http://www.weedscience.org.

Iqbal, J., Wright, D., 1997. Effects of nitrogen supply on competition between wheat and three weed species. Weed Res. 37, 391–400.

Jabran, K., Chauhan, B.S., 2015. Weed management in aerobic rice systems. Crop. Prot. 78, 151–163.

Jamshidia, K., Yousefia, A.R., Oveisib, M., 2013. Effect of cowpea (*Vigna unguiculata*) intercropping on weed biomass and maize (*Zea mays*) yield. N. Z. J. Crop. Hortic. Sci. 41, 180–188.

Johnson, G.A., Hoverstad, T.R., 2002. Effect of row spacing and herbicide application timing on weed control and grain yield in corn (*Zea mays*). Weed Technol. 16, 548–553.

Johnson, G.A., Hoverstad, T.R., Greenwald, R.E., 1998. Integrated weed management using narrow row spacing, herbicides, and cultivation. Agron. J. 90, 40–46.

Jordan, N., 1996. Weed prevention: priority research for alternative weed management. J. Prod. Agric. 9, 485–490.

Jornsgard, B., Rasmussen, K., Hill, J., Christiansen, J.L., 1996. Influence of nitrogen on competition between cereals and their natural weed populations. Weed Res. 36, 461–470.

Juskiw, P.E., Helm, J.H., Salmon, D.F., 2002. Competitive ability in mixtures of small grain cereals. Crop Sci. 40, 159–164.

Khan, Z.R., Pickett, J.A., Wadhams, L.J., Hassanali, A., Midega, C.A.O., 2006. Combined control of *Striga hermonthica* and stemborers by maize-*Desmodium* spp. intercrops. Crop Prot 25, 989–995.

Khan, Z., Midega, C.A.O., Hooper, A., Pickett, J., 2016. Push-pull: chemical ecology-based integrated pest management technology. J. Chem. Ecol. 42, 689–697.

Korres, N.E., 2005. Encyclopaedic Dictionary of Weed Science: Theory and Digest. Lavoisier SAS; Intercept Ltd., France; UK, p. 695.

Korres, N.E., Froud-Williams, R.J., 2002. Effects of winter wheat cultivars and seed rate on the biological characteristics of naturally occurring weed flora. Weed Res. 42, 417–428.

Korres, N.E., Froud-Williams, R.J., 2004. The interrelationships of winter wheat cultivars, crop density and competition of naturally occurring weed flora. Biol. Agric. Hortic. 22, 1–20.

Korres, N.E., Norsworthy, J.K., 2015a. Influence of a rye cover crop on the critical period for weed control in cotton. Weed Sci. 631, 346–352.

Korres, N.E., Norsworthy, J.K., 2015b. In: Influence of Palmer amaranth density and emergence date on seed production in wide row and drill-seeded soybean. Proceedings of the Weed Science Society of America Annual Meeting, Lexington, KN.

Korres, N.E., Norsworthy, J.K., 2017. Palmer amaranth (*Amaranthus palmeri*) demographic and biological characteristics in wide-row soybean. Weed Sci. 65, 491–503.

Korres, N.E., Norsworthy, J.K., Brye, K.R., Skinner Jr, V., Mauromoustakos, A., 2017. Relationships between soil properties and the occurrence of the most agronomically important weed species in the field margins of eastern Arkansas. Implications on weed management. Weed Res. 57, 159–171.

Korres, N.E., Norsworthy, J.K., Tehranchian, P., Gitsopoulos, T.C., Loka, D.A., Oosterhuis, D.M., Moss, S., Gealy, D., Burgos, N.R., Miller, R., Palhano, M., 2016. Cultivars to face climate change effects on crops and weeds: a review. Agron. Sustain. Dev. 36, 12.

Koscelny, J.A., Peeper, T.F., Solie, J.B., Solomon, S.G., 1990. Effect of wheat (*Triticum aestivum*) row spacing, seeding rate and cultivar on yield loss from cheat (*Bromus secalinus*). Weed Technol. 4, 487–492.

Lamichhane, J.R., Devos, Y., Beckie, H.J., Owen, M.K.D., Tillie, P., Messean, A., Kudsk, P., 2017. Integrated weed management systems with herbicide-tolerant crops in the European Union: lessons learnt from home and abroad. Crit. Rev. Biotechnol. 37, 459–475.

Lemerle, D., Verbeek, B., Coombes, N., 1995. Losses in grain yield of winter crops from *Lolium rigidum* competition depend on crop species, cultivar and season. Weed Res. 35, 503–509.

Lemerle, D., Verbeek, B., Orchard, B., 2001. Ranking the ability of wheat varieties to compete with *Lolium rigidum*. Weed Res. 41, 197–209.

Liebl, R., Simmons, F.W., Wax, L.M., Stoller, E.W., 1992. Effects of rye (*Secale cereale*) mulch on weed control and soil moisture in soybean (*Glycine max*). Weed Technol. 6, 838–846.

Liebman, M., Davis, A.S., 2000. Integration of soil, crop and weed management in low-external-input farming systems. Weed Res. 40, 27–47.

Liebman, M., Dyck, E., 1993. Crop rotation and intercropping strategies for weed management. Ecol. Appl. 3, 92–122.

Liebman, M., Miller, Z.J., Williams, C.L., Westerman, P.R., Dixon, P.M., Heggenstaller, A., Davis, A.S., Menalled, F.D., Sundberg, D.N., 2014. Fates of *Setaria faberi* and *Abutilon theophrasti* seeds in three crop rotation systems. Weed Res. 54, 293–306.

Lintell-Smith, G., Baylis, J.M., Watkinson, A.R., Firbank, L.G., 1992. The effects of reduced nitrogen and weed competition on the yield of winter wheat. Asp. Appl. Biol. 30, 367–372.

Love, S.L., Eberlein, C.V., Stark, J.C., Bohl, W.H., 1995. Cultivar and seedpiece spacing effects on potato competitiveness with weeds. Am. Potato J. 72, 197–213.

Lutman, P.J.W., 2018. Sustainable weed control in oilseed rape. In: Korres, N.E., Burgos, N.R., Duke, S.O. (Eds.), Weed Control. Sustainability, Hazards and Risks in Cropping Systems Worldwide. CRC Press, Boca Raton, FL. ISBN: 978-1498787468 (in press).

Maqbool, M.M., Tanveer, A., Ali, A., Abbas, M.N., Imran, M., Ahmad, M., Abid, A.A., 2016. Growth and yield response of maize (*Zea mays*) to inter and intra-row weed competition under different fertilizer application methods. Planta Daninha 34, 47–56.

Malik, V.S., Swanton, C.J., Michaels, T.E., 1993. Interaction of white bean (*Phaseolus vulgaris* L.) cultivars, row spacing and seeding density with annual weeds. Weed Sci 41, 62–68.

Marin, C., Weiner, J., 2014. Effects of density and sowing pattern on weed suppression and grain yield in three varieties of maize under high weed pressure. Weed Res. 54, 467–474.

Matiello, J.B., Santinato, F., 2016. Corda-de-viola avança nos cafezais. Folha Tecnica numero 318Fundaçao Procafe, Varginha, Brasil.

Midega, C.A.O., Salifu, D., Bruce, T.J., Pittchar, J., Pickett, J.A., Kahn, Z.R., 2014. Cumulative effects and economic benefits of intercropping maize with food legumes on *Striga hermonthica* infestations. Field Crop Res. 155, 144–152.

Mohler, C.L., 1996. Ecological bases for the cultural control of annual weeds. J. Prod. Agric. 9, 468–474.

Mortensen, D.A., Egan, J.F., Maxwell, B.D., Ryan, M.R., Smith, R.G., 2012. Navigating a critical juncture for sustainable weed management. Bioscience 62, 75–84.

Norsworthy, J.K., Korres, N.E., Walsh, M.J., Powles, S.B., 2016. Integrating herbicide programs with harvest weed seed control and other fall management practices for the control of glyphosate-resistant Palmer amaranth. Weed Sci. 64, 540–550.

Okafor, L.I., De Datta, S.K., 1976. Competition between upland rice and purple nutsedge for nitrogen, moisture and light. Weed Sci. 24, 43–46.

Olesen, J.E., Hansen, P.K., Berntsen, J., Christensen, S., 2004. Simulation of above-ground suppression of competing species and competition tolerance in winter wheat varieties. Field Crop Res. 89, 263–280.

Owen, M.D.K., Beckie, H.J., Leeson, J.Y., Norsworthy, J.K., Steckel, L.E., 2015. Integrated pest management and weed management in the United States and Canada. Pest Manag. Sci. 71, 357–376.

Price, A.J., Balkcom, K.S., Raper, R.L., Monks, C.D., Baventine, R.M., Iversen, K.V., 2008. Controlling Glyphosate-Resistant Pigweed in Conservation Tillage Cotton Systems. USDAARS-NSDL, Auburn, AL. Special Publication 09.

Price, A.J., Norsworthy, J.K., 2013. Cover crops for weed management in southern reduced tillage vegetable cropping systems. Weed Technol. 27, 212–217.

Radford, B.J., Wilson, B.J., Cartledge, O., Wakins, F.B., 1980. Effect of wheat seeding rate on wild oat competition. Aust. J. Exp. Agric. Anim. Hus. 20, 77–81.

Rajcan, I., Swanton, C.J., 2001. Understanding weed competition: resource competition, light quality and the whole plant. Field Crop Res. 71, 139–150.

Riar, D.S., Norsworthy, J.K., Steckel, L.E., Stephenson IV, D.O., Eubank, T.W., Bond, J., Scott, R.C., 2013. Adoption of best management practices for herbicide-resistant weeds in midsouthern United States cotton, rice, and soybean. Weed Technol. 27, 788–797.

Saini, M., Price, A.J., van Santen, E., 2006. In: Cover crop residue effects on early-season weed establishment in a conservation tillage corn-cotton rotation. Proceedings of the 28th Southern Conservation Systems Conference. Bushland, TX: USDA-ARS Conservation and Production Research Laboratory Report No. 06-1, pp. 175–178.

Sardana, V., Mahajan, G., Jabran, K., Chauhan, B.S., 2017. Role of competition in managing weeds: an introduction to the special issue. Crop. Prot. 95, 1–7.

Schomberg, H.H., McDaniel, R.G., Mallard, E., Endale, D.M., Fisher, D.S., Cabrera, M.L., 2006. Conservation tillage and cover crop influences on cotton production on a southeastern U.S. coastal plain soil. Agron. J. 98, 1247–1256.

Scott, R.K., Wilcockson, S.J., 1976. Weed biology and their growth in sugar beet. Ann. Appl. Biol. 83, 331–335.

Simic, M., Spasojevic, I., Kovacevic, D., Brankov, M., Dragievic, V., 2016. Crop rotation influence on annual and perennial weed control and maize productivity. Rom. Agric. Res. 33, 125–132.

Simko, I., Hayes, R.J., Mou, B., McCreight, J.D., 2014. Lettuce and spinach. In: Smith, S., Diers, B., Specht, J., Carver, B. (Eds.), Yield Gains in Major U.S. Field Crops. In: 33, CSSA Special Publication, ASA, CSSA, and SSSA, Madison, WI, pp. 53–85.

Skroch, W.A., Powell, M.A., Bilderback, T.E., Henry, P.H., 1992. Mulches: durability, aesthetic value, weed control and temperature. J. Environ. Hortic. 10, 43–45.

Smith, J., 1980. The relevance of plant population. Proceedings of 16th NIAB Conference, pp. 41–42.

Song, L., Zhang, D.W., Li, F.M., Fan, X.W., Ma, Q., Turner, N.C., 2010. Soil water availability alters the inter- and intra-cultivar competition of three spring wheat cultivars bred in different eras. J. Agron. Crop Sci. 196, 323–335.

Soto-Pinto, L., Perfecto, I., Caballero-Nieto, J., 2002. Shade over coffee: its effects on berry borer, leaf rust and spontaneous herbs in Chiapas, Mexico. Agrofor. Syst. 55, 37–45.

Stevanato, P., Trebbi, D., Bertaggia, M., Colombo, M., Broccanello, C., Concheri, G., Saccomani, M., 2011. Root traits and competitiveness against weeds in sugar beet. Int. Sugar J. 113, 497–501.

Strek, H.J., 2014. Herbicide resistance. What have we learned from other disciplines? J. Chem. Biol. 7, 129–132.

Stumm, C., Berg, M., Kopke, U., 2009. In: Mayer, J., Alföldi, T., Leiber, F., Dubois, D., Fried, P., Heckendorn, F. et al., (Eds.), Anbau und Düngung von Winterraps (Brassica napus L.) im Okologischen Landbau. Pflanzenbau: Tagungsbandes der 10. Wissenschaftstagung Ökologischer Landbau. In: Band 1, pp. 193–196. Zurich Feb. 2009.

Stupnicka-Rodzynkiewicz, E., Labza, T., Hochol, T., 1999. Nutrient accumulation by weeds as an indicator of their competitiveness against crop plants. Proceedings of 11 European Weed Research Society Symposium, p. 22.

Teasdale, J.R., 1995. Influence of narrow row high population corn (Zea mays) on weed-control and light transmittance. Weed Technol. 9, 113–118.

Teasdale, J.R., 1998. Influence of corn (Zea mays) population and row spacing on corn and velvetleaf (Abutilon theophrasti) yield. Weed Sci. 46, 447–453.

Teasdale, J.R., Mohler, C.L., 1993. Light transmittance, soil temperature, and soil moisture under residue of hairy vetch and rye. Agron. J. 85, 673–680.

Teyker, R.H., Hoelzer, H.D., Liebl, R.A., 1991. Maize and pigweed response to nitrogen supply and form. Plant Soil 135, 287–292.

Thompson, H., 2012. War on weeds loses ground. Nature 485, 430. https://doi.org/10.1038/485430a.

Tollenaar, M., Dibo, A.A., Aquilera, A., Weise, S.F., Swanton, C.J., 1994. Effect of crop density on weed interference in maize. Agron. J. 86, 591–595.

Vandermeer, J., 1989. The Ecology of Intercropping. Cambridge University Press, Cambridge, UK.

Vencill, W.K., Nichols, R.L., Webster, T.M., Soteres, J.K., Smith, C.M., Burgos, N.R., Johnson, W.G., McClelland, M.R., 2012. Herbicide resistance: toward an understanding of resistance development and the impact of herbicide-resistant crops. Weed Sci. 60, 2–30.

Vidotto, F., Fogliatto, S., Milan, M., Ferrero, A., 2016. Weed communities in Italian maize fields as affected by pedoclimatic traits and sowing time. Eur. J. Agron. 74, 38–46.

Vollmann, J., Wagentristl, H., Hartl, W., 2010. The effects of simulated weed pressure on early maturity soybeans. Eur. J. Agron. 32, 243–248.

Watson, G.W., 1988. Organic mulch and grass competition influence tree root development. J. Arboric. 14, 200–203.

Weed Ecology and Management Laboratory, 2017. Cornell University. https://weedecology.css.cornell.edu/manage/manage.php?id=8.

Williams II, M.M., Boydston, R.A., 2013. Crop seeding level: implications for weed management in sweet corn. Weed Sci. 61, 437–442.

Wilson, P.J., 1999. The effect of nitrogen on populations of rare arable plants in Britain. In field margins and buffer zones: ecology, management and policy. Asp. Appl. Biol. 54, 93–100.

Woomer, P.L., Bokanga, M., Odhiambo, G.D., 2008. Striga management and the African farmer. Outlook Agric. 37, 277–282.

Wortman, S.E., Davis, A.S., Schutte, B.J., Lindquist, J.L., 2011. Integrating management of soil nitrogen and weeds. Weed Sci. 59, 162–170.

Wright, K.J., 1993. In: Weed seed production as affected by crop density and nitrogen application.Brighton Crop Protection Conference-Weeds, vol. 1, pp. 275–280.

Zhang, J., Bartholomew, D.P., 1997. Effect of plant population density on growth and dry-matter partitioning of pineapple. Acta Hortic. 425, 363–376.

Biological Weed Control

Ahmet Uludag*,†, Ilhan Uremis‡, Mehmet Arslan§

*Çanakkale Onsekiz Mart University, Çanakkale, Turkey †Düzce University, Düzce, Turkey
‡Mustafa Kemal University, Hatay, Turkey §Erciyes University, Kayseri, Turkey

7.1 INTRODUCTION

The biological weed control methods use living organisms in plant protection and have been defined as "the use of an agent, a complex of agents, or biological processes to bring about weed suppression" (WSSA, 2017). Biological weed control has several advantages compared with other weed control methods. These benefits may include a reduced contamination of soil, water, and food by herbicide residues. There are many reasons for using biological control in weed science such as loss of many common herbicides due to problems such as tight regulations or evolution of herbicide resistance in weeds; changing understanding in weed control such as targeting only unwanted species, conserving environmentally sensitive or prone to degradation areas, avoiding contamination due to chemicals; and inclination to healthier and sustainable cropping systems. Furthermore, biological control is considered cheaper and self-sufficient if the organisms released get establish successfully and reproduce. However, invasiveness of some agents and the effects on nontarget organisms have been reported (Myers and Cory, 2017; Jones et al., 2017; Van Lenteren, 2012; Weyl and Martin, 2016; Van Wilgen et al., 2013), suggesting caution when choosing and releasing biological agents in weed control.

Many different creatures have been used or suggested for biological weed control ranging from microscopic rhizobacteria to large mammals. For example, *Ctenopharyngodon idella* Val., a fish, is used to control aquatic weeds (Domingues et al., 2016). Use of goats has also been suggested (Chalak and Pannell, 2015; Khan et al., 2009). However, these animals will not be discussed in this chapter. Use of arthropods, mainly insects and bioherbicides, will be covered.

Non-Chemical Weed Control
https://doi.org/10.1016/B978-0-12-809881-3.00007-3

7.2 USE OF INSECTS FEEDING ON THE WEEDS

The known history of using insects feeding on weeds goes back to the 19th century. *Dactylopius ceylonicus* (Green) (Hemiptera, Dactylopiidae) was brought to India to produce cochineal dye. This importation led to the biological control of *Opuntia monacantha* (Willdenow) Haworth (Cactaceae) instead of producing dye successfully. Following India, the insect was released in South Africa and Australia where it provided a successful weed control between 1796 and 1809 (Van Wilgen et al., 2013). Many early releases can be found in the literature to control *Opuntia* spp. using different insects (Winston et al., 2014). Biological controls of *Lantana camara* L. and *Hypericum perforatum* L. are other successful examples from the early history of using insects for weed management. A comprehensive list of released insects on given weeds was provided by Winston et al. (2014). Table 7.1 provides list of weeds controlled by insects with or without other biological techniques.

Three species, namely, *Parthenium hysterophorus* L., *Convolvulus arvensis* L., and *Solanum elaeagnifolium* Cav. were chosen to give some insights on the use of insects in biological control. Biological control programs with insects have been suggested or applied to control *P. hysterophorus* in all countries that have been widely invaded by this weed. Mites have also been used for biological control of *C. arvensis*, which are included in this chapter because they are arthropods too. *S. elaeagnifolium* is a weed in many countries, but its biological control is not common; only South Africa has a successful program.

7.2.1 Parthenium hysterophorus

Parthenium hysterophorus is an invasive plant, which was originated in Americas and currently invades croplands and noncropped areas with different climatic, edaphic, and geographic conditions in more than 40 countries on five continents (Adkins and Shabbir, 2014). The biological control of *P. hysterophorus* has been stated as successful in Australia where the first biological control program for *P. hysterophorus* management was initiated in 1977 and considered as the most successful country controlling *P. hysterophorus* (Dhileepan and Mcfayden, 2012; Terblanche et al., 2016).

Zygogramma bicolorata Pallister (Chrysomelidae, Coleoptera) that has Mexican origin is an effective organism to control *P. hysterophorus*. Both larvae and adults of this insect feed on

TABLE 7.1 The Most Successful Releases of Arthropods Feeding on Plants

Weeds			Agents		
Species	Family	Origin	Species	Order, Family	Country Released
Alternanthera philoxeroides (Mart.) (Griseb.)	Amaranthaceae	South America	*Agasicles hygrophila* (Selman and Vogt)	Coleoptera, Chrysomelidae	Australia, 1977
					New Zealand, 1982
					China, 1986
					Puerto Rico, 1997
					Thailand, 1981
					The United States, 1964

TABLE 7.1 The Most Successful Releases of Arthropods Feeding on Plants—cont'd

Weeds			Agents		
Species	Family	Origin	Species	Order, Family	Country Released
Pistia stratiotes L.	Araceae	Tropical Americas, Asia, Malaysia, Australia (NT)	*Neohydronomus affinis* (Hustache)	Coleoptera, Curculionidae	Australia, 1982
					Benin, 1995
					Botswana, 1987
					Cote d'Ivoire, 1998
					Ghana, 1996
					Kenya, 1999
					Nigeria, 1997
					Papua New Guinea
					Puerto Rico, 1998
					Congo, 1999
					South Africa, 1985
					Togo, 2001
					Senegal, 1994
					The United States, 1987
					Vanuatu, 2006
					Zambia, 1991
					Zimbabwe, 1988
Ambrosia artemisiifolia L.	Asteraceae	North America	*Epiblema strenuana* (Walker)	Lepidoptera, Tortricidae	Australia, 1984
					China (continued)
Carduus nutans L. subsp. nutans	Asteraceae	Europe, Asia, northern Africa	*Rhinocyllus conicus* (Frölich)	Coleoptera, Curculionidae	Australia, 1988
					Australia, 1989
					New Zealand, 1973
C. nutans	Asteraceae	Europe, Asia, northern Africa	*Trichosirocalus mortadelo* (Alonso-Zarazaga and Sanchez-Ruiz)	Coleoptera, Curculionidae	Australia, 1993
C. pycnocephalus L.	Asteraceae	Europe, Asia, northern Africa	Cheilosia grossa (Fallén)	Diptera, Syrphidae	The United States, 1993
Centaurea cyanus L.	Asteraceae	Eurasia	*Chaetorellia australis* (Héring)	Diptera, Tephritidae	The United States, 1988

Continued

TABLE 7.1   The Most Successful Releases of Arthropods Feeding on Plants—cont'd

Weeds			Agents		
Species	Family	Origin	Species	Order, Family	Country Released
C. diffusa Lam.	Asteraceae	Eurasia	Larinus minutus (Gyllenhal)	Coleoptera, Curculionidae	The United States, 1991
C. stoebe L. sens. Lat.	Asteraceae	Eurasia	Cyphocleonus achates (Fåhraeus)	Coleoptera, Curculionidae	Canada, 1987
Chondrilla juncea L.	Asteraceae	Eurasia	Cystiphora schmidti (Rübsaamen)	Diptera, Cecidomyiidae	Australia, 1971
Chromolaena odorata (L.) (R. M. King and H. Rob.)	Asteraceae	Caribbean, tropical, and subtropical Americas	Actinote anteas (Doubleday)	Lepidoptera, Nymphalidae	Indonesia, 1999
Cirsium arvense (L.) (Scop.)	Asteraceae	Eurasia	R. conicus (Frölich)	Coleoptera, Curculionidae	Canada, 1968
Mikania micrantha (Kunth)	Asteraceae	Central America, South America	A. anteas (Doubleday)	Lepidoptera, Nymphalidae	Indonesia, 1999
Onopordum spp.	Asteraceae	Eurasia, northern Africa	L. latus (Herbst)	Coleoptera, Curculionidae	Australia, 1992
Tripleurospermum inodorum (L.) Sch. Bip.	Asteraceae	Eurasia	Omphalapion hookerorum (Kirby)	Coleoptera, Brentidae	Canada, 1992
Azolla filiculoides Lam.	Azollaceae	North America, Central America, South America	Stenopelmus rufinasus (Gyllenhal)	Coleoptera, Erirhinidae	South Africa, 1997; Zimbabwe, 1999
Cordia curassavica (Jacq.) (Roem. and Schult.)	Boraginaceae	South and Central America, Caribbean	Eurytoma attiva (Burks)	Hymenoptera, Eurytomidae	Malaysia, 1977
Cynoglossum officinale L.	Boraginaceae	Eurasia	Mogulones crucifer (Pallas)	Coleoptera, Curculionidae	Canada, 1997
Cereus jamacaru DC. subsp. jamacaru.	Cactaceae	South America	Hypogeococcus festerianus (Lizer y Trelles)	Hemiptera, Pseudococcidae	South Africa, 1983
O. aurantiaca Lindl.	Cactaceae	Argentina, Uruguay	Cactoblastis cactorum (Berg)	Lepidoptera, Pyralidae	Australia, 1926; South Africa, 1933
			D. austrinus (De Lotto)	Hemiptera, Dactylopiidae	Australia, 1933
O. ficus-indica (L.) (Mill.)	Cactaceae	Mexico	C. cactorum (Berg)	Lepidoptera, Pyralidae	Hawaii, the United States, 1950; South Africa, 1933

TABLE 7.1   The Most Successful Releases of Arthropods Feeding on Plants—cont'd

Weeds			Agents		
Species	Family	Origin	Species	Order, Family	Country Released
Acacia dealbata Link	Fabaceae	Australia	*Melanterius maculatus* Lea	Coleoptera, Curculionidae	South Africa, 1994
A. longifolia (Andrews) (Willd.)	Fabaceae	Australia	*M. ventralis* Lea	Coleoptera, Curculionidae	South Africa, 1985
Mimosa diplotricha (C. Wright)	Fabaceae	Tropical Americas	*Heteropsylla spinulosa* (Muddiman, Hodkinson, and Hollis)	Hemiptera, Psyllidae	American Samoa, 1997
Prosopis juliflora (Sw.) DC.	Fabaceae	Colombia, Ecuador, Mexico, Peru, Venezuela	*Algarobius prosopis* (Le Conte)	Coleoptera, Chrysomelidae	Ascension Island, 1997
Sesbania punicea (Cav.) (Benth.)	Fabaceae	South America	*Neodiplogrammus quadrivittatus* (Olivier)	Coleoptera, Curculionidae	South Africa, 1984
Myriophyllum aquaticum (Vell.) (Verdc.)	Haloragaceae	South America	*Lysathia* sp.	Coleoptera, Chrysomelidae	South Africa, 1994
Hypericum perforatum L.	Hypericaceae	Asia Minor, Europe, northern Africa	*Chrysolina quadrigemina* (Suffrian)	Coleoptera, Chrysomelidae	Hawaii, the United States, 1965 New Zealand, 1965
Lythrum salicaria L.	Lythraceae	Europe, northern Africa, Asia	*Galerucella calmariensis* (L.)	Coleoptera, Chrysomelidae	Canada, 1992
Orobanche minor Sm.	Orobanchaceae	Eurasia	*Phytomyza orobanchia* (Kaltenbach)	Diptera, Agromyzidae	Chile, 1998
Phelipanche ramosa (L.) (Pomel)	Orobanchaceae	Eurasia	*P. orobanchia* (Kaltenbach)	Diptera, Agromyzidae	Chile, 1998
Rumex spp.	Polygonaceae	Europe, Asia, northern Africa	*Pyropteron doryliformis* (Ochsenheimer)	Lepidoptera, Sesiidae	Australia, 1989
Eichhornia crassipes (Mart.) Solms	Pontederiaceae	South America	*Neochetina bruchi* (Hustache)	Coleoptera, Erirhinidae	Australia, 1990
E. crassipes	Pontederiaceae	South America	*Neochetina eichhorniae* (Warner)	Coleoptera, Erirhinidae	Australia, 1975
Lantana camara L.	Verbenaceae	Tropical Americas	*Aceria lantanae* (Cook)	Acari, Eriophyidae	South Africa, 2007

Continued

7. BIOLOGICAL WEED CONTROL

TABLE 7.1 The Most Successful Releases of Arthropods Feeding on Plants—cont'd

Weeds			Agents		
Species	**Family**	**Origin**	**Species**	**Order, Family**	**Country Released**
L. camara	Verbenaceae	Tropical Americas	*Aconophora compressa* (Walker)	Hemiptera, Membracidae	Australia, 1995
L. camara	Verbenaceae	Tropical Americas	*Hypena laceratalis* (Walker)	Lepidoptera, Erebidae	Hawaii, the United States, 1957
Tribulus terrestris L.	Zygophyllaceae	Mediterranean, western Asia, Africa	*Microlarinus lareynii* (Jacquelin du Val)	Coleoptera, Curculionidae	Hawaii, the United States, 1967

Source: Winston, R.L., Schwarzländer, M., Hinz, H.L., Day, M.D., Cock, M.J.W., Julien, M.H., 2014. Biological Control of Weeds: A World Catalogue of Agents and Their Target Weeds, 5th ed. USDA Forest Service, Forest Health Technology Enterprise Team, Morgantown, West Virginia. FHTET-2014-04, 838 pp.

leaves during early growth stages of *P. hysterophorus* (Dhileepan et al., 2000). The beetle was released deliberately during 1980 in Australia, 1984 in India, 2004 in Srilanka, and undeliberately during 2007 in Pakistan and 2009 in Nepal (Winston et al., 2014; McFadyen and McClay, 1981; Jayanth, 1987). Furthermore, its use for *P. hysterophorus* biocontrol in South Africa and Ethiopia was mentioned by Shabbir et al. (2015). Increase in the beetle population and visible effects were seen after 3 years of release in India and 10 years in Australia (Jayanth and Bali, 1994; Dhileepan et al., 2000). Feeding of beetle on sunflower leaves has been noted probably due to deposition of *P. hysterophorus* pollens, which might attract the beetle (Tanveer et al., 2015).

Epiblema. strenuana Walker (Lepidoptera, Tortricidae) is another successful arthropod with Mexican origin. This biological agent was released in Australia during 1982 to control *P. hysterophorus* (Callander and Dhileepan, 2016; Winston et al., 2014). It is considered a major contributor to substantial control of *P. hysterophorus* because of its high abundance and heavy impact especially when plants are young (Shabbir et al., 2015).

Several other insect and a rust species (*Puccinia abrupta* Dietel and Holw. var. *partheniicola*, (H. S. Jacks.) Parmelee) have been released in Australia that provided very good weed control (Dhileepan and McFadyen, 1997; Winston et al., 2014).

In Australia, *Stobaera concinna* and *Z. bicolorata* had populations established on *A. artemisiifolia*, which is another invasive alien plant in agricultural and nonagricultural areas (Winston et al., 2014). Mealybugs were reported feeding on *P. hysterophorus* and some other weeds (Javaid et al., 2006), but there has been no further report if they have been used deliberately. Australian biological control program on *P. hysterophorus* included two rust fungi with Mexican origin as well: *P. abrupta* var. *partheniicola* and *P. xanthii* Schwein. var. *parthenii-hysterophorae* Seier, H.C. Evans and Á. Romero. These were released in 1991 and 2000, respectively. The latter was brought from Texas, the United States, and Mexico. Successful results have been obtained with these agents. The latter was released in South Africa in 2010, which was brought from Australia. Also, the former fungi was recorded in Maurius during 1967, in India 1994, in Ethiopia and Kenya 1997, in South Africa 1995, in China 2002, and in Nepal 2011 although it was not released deliberately.

7.2.2 Convolvulus arvensis

Convolvulus arvensis is a perennial weed with Eurasian origin and is a problem globally both in agricultural and nonagricultural areas. The main agent used to control this weed was not an insect species but another arthropod, an acarian species, *A. malherbae* Nuzzaci (Eriophyidae), released in Canada and the United States in 1989, South Africa in 1994, and Mexico in 2004. The agent was established in North America but not in Mexico and South Africa (Winston et al., 2014). It is mainly effective in the North America, especially in dry environments although variations have been reported (Smith et al., 2010). However, successful colonization by the mite was reported in the highly humid areas in Missouri after 2010 release (Lauriault et al., 2013). *Calystegia sepium* (L.) R. Br was also targeted, but establishment and impact results are not clearly known (Winston et al., 2014). There was some concern about native *Calystegia* and *Convolvulus* species, which restricted it to be released California earlier. However, there has been no apparent damage on nontarget species in the states where it was released (Hinz et al., 2014). Also, an exotic lepidopteran species has been released, that is, *Tyta luctuosa* (Denis and Schiffermüller) in the United States and Canada during 1987 and 1989, respectively, but the establishment and impact especially in Canada was not clear (Winston et al., 2014).

In Europe, there has been research to find out agents to control *C. arvensis*. The larvae of *Melanagromyza albocilia* Hendel (Agromyzidae) and the larvae and adults of *Hypocassida subferruginea* (Schrank) (Chrysomelidae) were reported as the top candidates in Slovakia (Tóth et al., 2008). Actually, 140 insects, three mites, and three fungi were reported on *C. arvensis* in Mediterranean Europe (Rosenthal and Buckingham, 1982) that might be used as a source of agents to control this weed in its native habitat.

7.2.3 Solanum elaeagnifolium

Solanum elaeagnifolium is an important invasive alien plant in the introduced ranges in five continents and is a weed in its native range, that is, northeast of Central America and southwest of North America (EPPO, 2007; Boyd et al., 1984). It has been established in semiarid steppes, temperate and subtropical deserts, and Mediterranean climate zones although it is native to humid subtropical zones in Americas (Brunel, 2011; Uludag et al., 2016). This perennial plant affects crop and animal production, biodiversity, environment, and life quality and may invade areas such as Balkans, Central Europe, and Asia due to effect of climate change (Uludag et al., 2016).

In weed's native range, a native nematode, *Ditylenchus phyllobius* (Thorne) was redistributed in 1974, which resulted in decreased biomass and density of *S. elaeagnifolium* (Winston et al., 2014). This nematode has been recorded in Argentina and India, and it is speculated that it reached India accidentally with plants (Olckers and Zimmermann, 1991; EPPO, 2007).

Successful biological control of *S. elaeagnifolium* has been implemented in South Africa, although *D. phyllobius* has been neither rejected nor released, four of tested insects were rejected, and the others such as *Frumenta* spp. have been released (Klein, 2011). *Frumenta nephelomicta* Meyrick (Lepidoptera, Gelechiidae), an insect with Mexican origin, was released in South Africa in 1978, 1984, and 1985, but could not established due to small size of release at earlier releases and drought at the last one. Another *Frumenta* species with Texas origin released in 1989 was not able to establish, this time due to parasitism (Olckers, 1995). Two *Leptinotarsa* species, that is, *L. defecta* (Stål) and *L. texana* Schaeffer (Coleoptera, Chrysomelidae) from Texas were released in South Africa in 1992; these species got established, and the latter became very successful agent (Hoffmann et al., 1998; Winston et al., 2014). It was concluded in a preintroduction study that threat from imported agents were not more than native insects and imported agents could be released (Olckers and Hulley, 1995) and native species were not able to control introduced *S. elaeagnifolium* (Hill et al., 1993). After two decades, there was decision for release despite risk was a correct one (Sheppard et al., 2006). In spite of South Africa's successful biological control program, there has not been any release in Mediterranean Basin or Australia where an important problem exists (Uludag et al., 2016; Australian Weeds Committee, 2012; Heap, 2014).

7.3 BIOHERBICIDES

Weed control in modern agriculture has been relied on herbicides, but almost no new herbicide mode of actions has been explored. Bioherbicides developed from higher plants, microorganisms, or microbiological phytotoxins (Lamberth, 2016; Cai and Gu, 2016) have been used for controlling weeds in agricultural systems (Cordeau et al., 2016; Dayan et al., 2012). Generally, it has been considered that bioherbicides cannot act as substitute to synthetic herbicides, but they can be a supplementary tool in weed control (Boyette et al., 2008). According to Pacanoski (2015), bioherbicides have some advantages such as the following:

High-level selectivity.

Low side effects on nontarget organisms.

Hardly have residue problems.

No resistance problems so far.

The idea of using microbial agents in weed control goes back as early as the beginning of the 1900s, but their early usage started after the World War II (Pacanoski, 2015). Use of *Fusarium oxysporum* Schlet. to control *O. ficus-indica* in Hawaii is one of the earliest examples (1940) (Rana and Rana, 2016). The mass production of *Alternaria cuscutacidae* was used in *Cuscuta* spp. control in Russia. Same parasitic weeds were controlled with another fungal agent *Colletotrichum gloeosporioides* f. sp. *cuscutaei* in China during 1963. LuBao was a commercial mycoherbicide that was developed during those times and is still under use (Rana and Rana, 2016). Attempts were initiated to find out bioherbicides to control *R.* spp. in the United States (Inman, 1971) and *Rubus* spp. in Chile (Oehrens, 1977) during the late 1960s. The following decade was full of projects to develop bioherbicides, and the 1980s was the decade where huge numbers of scientific articles were published on bioherbicides (Charudattan, 1991). Different biological weed control agents have been given in Table 7.2.

Formulated bioherbicides were produced and available in the US market during the 1980s; these were the following: Devine®, Collego®, Casst® followed by Dr. BioSedge®, BioMal®, Stumpout®, Biochon®, Camperico®, Woad Warrior®, Smolder®, and Myco-Tech® (see Table 7.3 for registered bioherbicides and basic information) (Aneja et al., 2013). Devine, the first bioherbicide registered was developed in Abbott Laboratories and produced from a facultative fungus, *Phytophthora palmivora* Butl. This fungus causes root rot of *Morrenia*

TABLE 7.2   Fungal, Viral, and Bacterial Agents That Have Potential Use in Weed Control

Pathogens or Agents	Weeds	References
A. cassiae	*Senna obtusifolia* (L.) H. S. Irwin and Barneby *S. occidentalis* (L.) Link *Crotalaria spectabilis* Roth	Boyette (1988), Charudattan et al. (1986), and Walker (1983)
A. destruens	*Cuscuta.* spp.	Simmons (1998)
A. eichhorniae	*Eichhornia crassipes (Mart.) Solms*	Shabana (2005)
A. helianthi	*Xanthium strumarium* L.	Abbas et al. (2004)
Amphobotrys ricini	Euphorbiaceae	Holcomb et al. (1989) and Whitney and Taber (1986)
Ascochyta caulina *Cercospora chenopodii* *C. dubia*	*Chenopodium album* L.	Scheepens and van Zon (1982)
Bipolaris setariae	*Eleusine indica* (L.) Gaertner	Hoagland et al. (2007)
C. caricis	*Cyperus esculentus* L.	Hoagland et al. (2007)
Cochliobolus lunatus	*Echinochloa crus-galli* (L.) P. Beauv.	Scheepens (1987)
C. coccodes, F. lateritium	*Abutilon theophrasti* Medik.	Hodgson et al. (1988) and Walker (1981)
C. dematium	Leguminosae	Cardina et al. (1988)
C. gloesporioides	Leguminosae, Malvaceae, Convolvulaceae (*C.* spp.)	Daniel et al. (1973) and Mortensen and Makowski (1997)
C. graminicola	Gramineae	Hoagland et al. (2007)
C. orbiculare	*X. spinosum*	Auld et al. (1988)
C. truncatum	*Sesbania exaltata* (Raf.) Rydb. ex A.W.Hill	Boyette (1991)
Dichotomophthora indica *D. portulacea*	*Portulaca oleracea* L.	Evans and Ellison (1988)
Exserohilum monoceras	*Echinochloa* spp.	Zhang and Watson (1997)
F. lateritium	*Sida spinosa* L. *Anoda cristata* (L.) Schltdl. *Potamogeton* spp.	Walker (1981), Walker (1981), and Bernhardt and Duniway (1986)
F. lateritium	*Ambrosia trifida* L.	Hoagland et al. (2007)
F. oxysporum	*Phelipanche ramosa* (L.) Pomel	Kohlschmid et al. (2009)
Myrothecium verrucaria	*S. obtusifolia* *Portulaca* spp. *Euphorbia* spp.	Walker and Tilley (1997) and Boyette et al. (2007)
Phoma chenopodicola	*C. album, Cirsium arvense* (L.) Scop., *Setaria viridis* (L.) P. Beauv., *Mercurialis annua* L.	Cimmino et al. (2013)

Continued

TABLE 7.2 Fungal, Viral, and Bacterial Agents That Have Potential Use in Weed Control—cont'd

Pathogens or Agents	Weeds	References
P. herbarum	*Taraxacum officinale (L.) Weber ex F.H. Wigg*	Neumann and Boland (1999)
P. macrostoma	*T. officinale*	Smith et al. (2015)
Phomopsis convolvulus	*Convolvulus arvensis L.*	Hoagland et al. (2007)
Phyllachora cyperi	*Cyperus rotundus L.*	Hoagland et al. (2007)
Pyricularia sp.	*Digitaria sanguinalis (L.) Scop.*	Hoagland et al. (2007)
P. grisea	*E. indica*	Figliola et al. (1988)
Pseudocercospora nigricans	*S. obtusifolia*	Hofmeister and Charudattan (1987)
Sclerotinia sclerotiorum	Multiple species	Brosten and Sands (1986)
S. minor	*T. officinale, Trifolium repens L., Plantago minor* Garsault	Riddle et al. (1991)
Septoria tritici f. sp. *avenae*	*Avena fatua L.*	Madariaga and Scharen (1985)
Sphacelotheca holci B. halepense B. sorghicola C. graminoicola	*Sorghum halepense (L.) Pers.*	Massion and Lindow (1986), Chiang et al. (1989), and Winder and Van Dyke (1989)
Pepino mosaic virus	*Solanum nigrum L.*	Kazinczi et al. (2006)
Araujia mosaic virus	*Araujia hortorum E. Fourn.*	Elliott et al. (2009)
Obuda pepper virus	*S. nigrum*	Kazinczi et al. (2006)
Pseudomonas syringae pv. *tagetis*	*C. arvense*	Johnson et al. (1996)
P. fluorescens strain BRG100	*S. viridis*	Quail et al. (2002)
P. fluorescens strain WH6	Multiple weeds	Banowetz et al. (2008)
P. fluorescens strain D7	*Bromus tectorum L.*	Kennedy et al. (1991)

Modified from Hoagland, R.E., Boyette, C.D., Weaver, M.A., Abbas, H.K., 2007. Bioherbicides: research and risks. Toxin Rev. 26, 313–342.

odorata (Hook and Arn) Lindl. and able to be effective for a long time by remaining in soil as saprophytic (TeBeest, 1990). Today, over 200 plant pathogens are candidates to be marketed as bioherbicides (Pacanoski, 2015). For example, *F. oxysporum* as a potential mycoherbicide to control *Striga* spp. had affected *Striga* emergence by 81.8%–94.3% depending on weed species (Marley et al., 2005). *P. fluorescens*, strain BRG100 had significantly affected the emergence and root development of a grass weed (Pedras et al., 2003).

TABLE 7.3 General Information on Registered Bioherbicides Worldwide

Country and Time	Products and Pathogens or Agents	Target Weed	Status	Formulation	References
The United States, 1960	*Acremonium diospyri*	*Diospyros virginiana* L. trees in rangelands	Unknown	Conidial suspension	Aneja et al. (2013)
China, 1963	LuBao 1, *Colletrichum gloeosporioides* f. sp. *cuscutae*	*Cuscuta* spp. in soybean	Available	Conidial suspension	Templeton (1992)
The United States, 1981	DeVine®, *Phytophtora palmivora*	*Morrenia odorata (Hook. and Arn.) Lindl.* in citrus orchard	Unknown	Liquid spores suspension	Ridings (1986) and Cordeau et al. (2016)
The United States, 1982/ 2006	Collego™/LockDown™, *C. gloeosporioides* f. sp. *aeschynomene*	*Aeschynomene virginica* (L.) Britton, Sterns, and Poggenb in soybean and rice	Not produced or distributed since 2003	Wettable powder	Bowers (1986), Smith (1991), and Cordeau et al. (2016)
The United States, 1983	Casst™, *A. cassiae*	*Cassia obtusifolia* L., *C. occidentalis* L., *C. spectabilis* DC. in soybean and peanut	No longer available	Solid	Charudattan et al. (1986)
The United States, 1984	ABG-5003, *C. rodmanii*	*Eichhornia crassipes (Mart.) Solms*	Unknown	Wettable powder	Charudattan (2001)
The United States, 1987	Dr. BioSedge®, *P. canaliculata*	*Cyperus esculentus* L. in soybean, potato, corn, and cotton	Product failed due to uneconomic production system	Emulsified suspension	Phatak (1992)
Canada, 1987	Velgo®, *C. coccodes*	*Abutilon theophrasti Medik.* in corn and soybean	Commercially available	Wettable powder	Butt (2000)
Canada, 1992	BioMal®, *C. gloeosporioides* f. sp. *malvae*	*Malva pusilla Sm.* in wheat, lentil, and flax	No longer available	Mallet/ wettable powder in silica gel	Boyetchko (1999) and Mortenson (1998)
South Africa, 1997	Stumpout™, *Cylindrobasidium laeve*	*Poa annua* L. in golf courses; *A. mearnsii* (De Wild) and *A. pycnantha* (Benth.) in native vegetations	Available	Liquid (oil) suspension	Shamoun and Hintz (1998)

Continued

TABLE 7.3   General Information on Registered Bioherbicides Worldwide—cont'd

Country and Time	Products and Pathogens or Agents	Target Weed	Status	Formulation	References
The Netherlands 1997	Biochon™, *Chondrostereum purpureum*	*Prunus serotina* (Ehrh.) in forests	No available since 2000	Mycelial suspension in water	Dumas et al. (1997)
Japan, 1997	Camperico®, *Xanthomonas campestris* pv. *poae*	*P. annua* in golf courses	No available	Suspension	Imaizumi et al. (1997)
The United States, 2002	Woad Warrior®, *P. thalaspeos*	*Isatis tinctoria* L. in farms, rangelands, and roadsides	Available but production limited	Powder	Stirk et al. (2006)
Canada, 2004/2007	Chontrol™ = Ecoclear™, *C.purpureum*	Hardwoods in forests	Commercially available	Spray emulsion and paste	Bailey et al. (2010)
Canada, 2004/2007	Myco-Tech™ paste, *C. purpureum*	Deciduous tree species in forests	Commercially available	Paste	Charudattan (2005)
Canada, 2007	Sarritor®, *S. minor*	*Taraxacum officinale (L.) Weber ex F.H. Wigg* in lawns, turf	Commercially available	Granular	Aneja et al. (2013)
The United States, 2008	Smolder®, *A. destruens*	*Cuscuta* spp. in fields and ornamental nursery	Registered	Conidial suspension	Aneja et al. (2013)
The United States, 2009	SolviNix™, tobacco mild green mosaic tobamovirus (TMGMV)	*Solanum viarum Dunal* in rangelands	Commercially available	Foliar spray suspension (liquid concentrate/ wettable powder)	Aneja et al. (2013)
Canada, 2010	Organo-Sol®, *Lactobacillus* spp.-fermented milk	*Trifolium, Medicago*, and *Oxalis* spp. in rights of way, forests	Registered	Liquid	Cordeau et al. (2016)

Modified from Rana, S.S., Rana, M.C., 2016. Principles and Practices of Weed Management. Department of Agronomy. College of Agriculture. CSK Himachal Pradesh Krishi Vishvavidyalaya, Palampur, p. 138.

S. viarum Dunal was affected by another bioherbicide candidate, that is, tobacco mild green mosaic virus (TMGMV) (Charudattan et al., 2003).

In spite of a high number of studies on bioherbicides, the number of commercial preparations introduced to market has been very small (Auld et al., 2003; Pacanoski, 2015). This has been attributed to several limiting factors such as environment (temperature and humidity), biology (host difference and resistance), and technology (formulation and mass production) (Auld et al., 2003). Due to the effect on sporulation performances of bioherbicides, humidity is an important environmental factor compared with temperatures (TeBeest et al., 1992). In addition, the C/N ratio has a significant effect on fungal sporulation (Jackson and Bothast, 1990). Another factor impacting the success of bioherbicides is the formulation since it is difficult to maintain a living organism to successfully reach on target plants under field conditions. Many organisms do not show good performance in vivo conditions as they showed in vitro studies (Cai and Gu, 2016). Research on new formulations requires time and finance. The registration procedure is another factor that negatively affects development of a new formulation according to legislations of any country (a period of 5 years may be required) (Auld et al., 2003). Despite many advantages, there are some risks/disadvantages of bioherbicides (Ash, 2010). A few of these are allergic problems to people who are exposed to them, spread of host plant, side effects on nontarget organisms especially beneficial microorganisms, and toxicity of metabolites on mammals (Hoagland et al., 2007). For instance, *P. melampodii*, a fungus with Mexican origin was applied to *P. hysterophorus* in Australia, subsequently affected *Tagetes* spp. and sunflower (Evans, 2000). Metabolites emitted by fungi not only are useful medicines or bioherbicides but also show high level of toxicity (fumonisins, ochratoxins, patulin, and zearalenone) or carcinogens (moniliformin and aflatoxin). However, there is not much information available on the effect of metabolites of fungal biocontrol agents, commercial mycoherbicides, mycoinsecticides, and possible risk of metabolites of saprophytic fungi contaminating foods. It has been suggested that mycoherbicides can be used safely to kill trunks of cut woods in forests and weed control in entertainment and sport areas (Abramson, 1998; Vey et al., 2001).

7.4 CONCLUSIONS

The biological weed control has a long history where animals, fungi, arthropods, or microorganisms have been used to feed upon, parasitize, or interfere with a targeted weed species. Australia, Canada, New Zealand, South Africa, and the United States were the first countries to apply biological approach for weed control. Indian subcontinent and China and some other Asian countries are applying biological weed control at increasing rate. But Europe has fewer biological control applications, which is attributed to public perceptions, funding, and legal constrains. Numerous biological control agents have been reported to reduce the incidence of many weeds. Biological agents are considered environment friendly and economically feasible way to control weeds, since they leaves no chemical residues that might have harmful effects on human societies or other organisms.

References

Abbas, H., Johnson, B., Pantone, D., Hines, R., 2004. Biological control and use of adjuvants against multiple seeded cocklebur (*Xanthium strumarium*) in comparison with several other cocklebur types. Biocontrol Sci. Technol. 14, 855–860.

Abramson, D.M., 1998. Mycotoxin formation and environmental factors. In: Sinha, K.K., Bhatnagar, D. (Eds.), Mycotoxins in Agriculture and Food Safety. Marcel Dekker, New York, pp. 255–277.

Adkins, S., Shabbir, A., 2014. Biology, ecology and management of the invasive parthenium weed (*Parthenium hysterophorus* L.). Pest Manag. Sci. 10 (7), 1023–1029.

Aneja, K.R., Kumar, V., Jiloha, P., Kaur, M., Sharma, C., Surain, P., Dhiman, R., Aneja, A., 2013. Potential bioherbicides: Indian perspectives. In: Salar, R.K., Gahlawat, S.K., Siwach, P., Duhan, J.S. (Eds.), Biotechnology: Prospects and Applications. Springer Science and Business Media, New Delhi, pp. 197–216.

Ash, G.J., 2010. The science, art and business of successful bioherbicides. Biol. Control 52, 230–240.

Auld, B.A., Mc Rae, C.F., Say, M.M., 1988. Possible of *Xanthium spinosum* by a fungus. Agric. Ecosyst. Environ. 21, 219–223.

Auld, B.A., Hertherington, S.D., Smith, H.E., 2003. Advances in bioherbicide formulation. Weed Biol. Manage. 3, 61–67.

Australian Weeds Committee, 2012. Weeds of National Significance: Silverleaf Nightshade (*Solanum elaeagnifolium*) Strategic Plan. Australian Weeds Committee, Canberra.

Bailey, K.L., Boyetchko, S.M., Langle, T., 2010. Social and economic drivers shaping the future of biological control: a Canadian perspective on the factors affecting the development and use of microbial biopesticides. Biol. Control 52, 221–229.

Banowetz, G.M., Azevedo, M.D., Armstrong, D.J., Halgren, A.B., Mills, D.I., 2008. Germination-arrest factor (GAF): biological properties of a novel, naturally-occurring herbicide produced by selected isolates of rhizosphere bacteria. Biol. Control 46, 380–390.

Bernhardt, E.A., Duniway, J.M., 1986. Decay of pondweed and *Hydrilla hibernacula* by fungi. J. Aquat. Plant Manage. 24, 20–24.

Bowers, R.C., 1986. Commercialization of Collego—an industrialist's view. Weed Sci. 34 (Suppl. 1), 24–25.

Boyd, J.W., Murray, D.S., Tyrl, R.J., 1984. Silverleaf nightshade, *Solanum elaeagnifolium*, origin, distribution, and relation to man. Econ. Bot. 38, 210–216.

Boyetchko, S.M., 1999. Innovative application of microbial agents for biological weed control. In: Mukerji, K.G., Chamola, B.P., Upadhyay, R.K. (Eds.), Biotechnological Approaches in Biocontrol of Plant Pathogens. Plenum Publishers, New York, pp. 73–98.

Boyette, C.D., 1988. Biocontrol of three leguminous weed species with *Alternaria cassiae*. Weed Technol. 2, 414–417.

Boyette, C.D., 1991. Host range and virulence of *Colletotrichum truncatum*, a potential mycoherbicide for hemp sesbania (*Sesbania exaltata*). Plant Dis. 75, 62–64.

Boyette, C.D., Hoagland, R.E., Weaver, M.A., 2007. Biocontrol efficacy of *Colletotrichum truncatum* for hemp sesbania (*Sesbania exaltata*) is enhanced with unrefined corn oil and surfactant. Weed Biol. Manage. 7, 70–76.

Boyette, C.D., Hoagland, R.E., Weaver, M.A., 2008. Interaction of a bioherbicide and glyphosate for controlling hemp sesbania in glyphosate-resistant soybean. Weed Biol. Manage. 8, 18–24.

Brosten, B.S., Sands, D.C., 1986. Field trials of *Sclerotinia sclerotiorum* to control Canada thistle (*Cirsium arvense*). Weed Sci. 34, 377–380.

Brunel, S., 2011. Pest risk analysis for *Solanum elaeagnifolium* and international management measures proposed. EPPO Bull. 41 (2), 232–242.

Butt, T.M., 2000. Fungal biocontrol agents. Pesticide Outlook October, 186–191. https://doi.org/10.1039/b008009h.

Cai, X., Gu, M., 2016. Bioherbicides in organic horticulture. Horticulturae 2, 3. https://doi.org/10.3390/horticulturae2020003.

Callander, J.T., Dhileepan, K., 2016. Biological control of parthenium weed: field collection and redistribution of established biological control agents. 20th Australasian Weeds Conference, Perth, Western Australia, 11–15 September 2016. Weeds Society of Western Australia, pp. 242–245.

Cardina, J., Littrell, R.H., Hanlin, R.T., 1988. Anthracnose of Florida beggarweed (*Desmodium tortuosum*) caused by *Collectotrichum truncatum*. Weed Sci. 36, 329–334.

Chalak, M., Pannell, D.J., 2015. Optimal integrated strategies to control an invasive weed. Can. J. Agric. Econ./Rev. can. d'agroecon. 63 (3), 381–407.

Charudattan, R., Elliot, M., Devalerio, J.T., Hiebert, E., Pettersen, M.E., 2003. In: Tobacco mild green mosaic virus: a virus-based bioherbicide. VI International Bioherbicide Group Workshop, Canberra, Australia.

Charudattan, R., 1991. The mycoherbicide approach with plant pathogens. In: TeBeest, D.O. (Ed.), Microbial Control of Weeds. Chapman & Hall, New York, pp. 24–57.

Charudattan, R., 2001. Biological control of weeds by means of plant pathogens: significance for integrated weed management in modern agroecology. BioControl 46, 229–260.

Charudattan, R., 2005. Ecological, practical, and political inputs into selection of weed targets: what makes a good biological control target? Biol. Control 35, 183–196.

Charudattan, R., Walker, H.L., Boyette, C.D., Ridings, W.H., TeBeest, D.O., Van Dyke, C.G., Worsham, A.D., 1986. Evaluation of *Alternaria cassiae* as a mycoherbicide for sicklepod (*Cassia obtusifolia*) in regional field tests. In: Southern Cooperative Service Bulletin. Alabama Agricultural Experiment Station, Auburn Univ, USA, pp. 1–19.

Chiang, M.Y., Leonard, K.J., Van Dyke, C.G., 1989. *Bipolaris halepense*: a new species from *Sorghum halepense* (johnsongrass). Mycologia 81, 532–538.

Cimmino, A., Andolfi, A., Zonno, M.C., Avolio, F., Santini, A., Tuzi, A., Berestetskyi, A., Vurro, M., Evidente, A., 2013. Chenopodolin: aphytotoxic unrearranged ent-pimaradiene diterpene produced by *Phoma chenopodicola*, a fungal pathogen for *Chenopodium album* biocontrol. J. Nat. Prod. 76, 1291–1297.

Cordeau, S., Triolet, M., Wayman, S., Steinberg, C., Guillemin, J.P., 2016. Bioherbicides: dead in the water? a review of the existing products for integrated weed management. Crop. Prot. 87, 44–49.

Daniel, J.T., Templeton, G.E., Smith Jr., R.J., Fox, W.T., 1973. Biological control of northern joinvetch in rice with an endemic fungal disease. Weed Sci. 21, 303–307.

Dayan, F.E., Owens, D.K., Duke, S.O., 2012. Rationale for a natural products approach to herbicide discovery. Pest Manag. Sci. 68 (4), 519–528.

Dhileepan, K., McFadyen, R.E., 1997. Biological control of parthenium in Australia—progress and prospects.Proceedings of the First International Conference on Parthenium Management, Vol. 1. pp. 40–44.

Dhileepan, K., Mcfayden, R.C., 2012. Current status of parthenium (*Parthenium hysterophorus* Linn.) biological control in Australia. J. Biol. Control. 26 (2), 31–36.

Dhileepan, K., Setter, S.D., McFadyen, R.E., 2000. Response of the weed *Parthenium hysterophorus* (Asteraceae) to defoliation by the introduced biocontrol agent *Zygogramma bicolorata* (Coleoptera: Chrysomelidae). Biol. Control 19, 9–16.

Domingues, F.D., Starling, F.L., Nova, C.C., Loureiro, B.R., Branco, C.W., 2016. The control of floating macrophytes by grass carp in net cages: experiments in two tropical hydroelectric reservoirs. Aquac. Res. 1–13. https://doi.org/10.1111/are.13163.

Dumas, M.T., Wood, J.E., Mitchell, E.G., Boyonoski, N.W., 1997. Control of stump sprouting of populus tremuloides and *P. grandidentata* by inoculation with *Chondrostereum purpureum*. Biol. Control 10, 37–41.

Elliott, M.S., Massey, B., Cui, X., Hiebert, E., Charudattan, R., Waipara, N., Hayes, L., 2009. Supplemental host range of Araujia mosaic virus, a potential biological control agent of moth plant in New Zealand. Australas. Plant Pathol. 38, 603–607.

EPPO, 2007. Solanum elaeagnifolium. Bull. OEPP/EPPO 37, 236–245.

Evans, H., Ellison, C., 1988. Preliminary work on the development of a mycoherbicide to control *Rottboellia cochinchinensis*. In: DelFosse, E.S. (Ed.), Proceedings of the VII International Symposium on Biological Control Weeds. Instituto Sperimentale per la Patologia Vegetale Ministero dell' Agricultura e delle Foreste (MAF), Rome, p. 76.

Evans, H.C., 2000. Evaluating plant pathogens for biological control of weeds: an alternative view of pest risk assessment. Australian. Plant Pathol. 29, 1–14.

Figliola, S.S., Camper, N.D., Ridings, W.H., 1988. Potential biological control agents for goosegrass (*Eleusine indica*). Weed Sci. 36, 830–835.

Heap, J.W., 2014. Silverleaf nightshade: progress and prospects for management of a new Australian WoNS. 19th Australasian Weeds Conference, "Science, Community and Food Security: the Weed Challenge," Hobart, Tasmania, Australia, 1–4 September 2014. Tasmanian Weed Society, pp. 166–169.

Hill, M.P., Hulley, P.E., Olckers, T., 1993. Insect herbivores on the exotic weeds *Solanum elaeagnifolium* Cavanilles and *S. sisymbriifolium* Lamarck (Solanaceae) in South Africa. Afr. Entomol. 1 (2), 175–182.

Hinz, H.L., Schwarzlander, M., Gassmann, A., Bourchier, R.S., 2014. Successes we may not have had: a retrospective analysis of selected weed biological control agents in the United States. Invasive Plant Sci. Manage. 7, 565–579.

Hoagland, R.E., Boyette, C.D., Weaver, M.A., Abbas, H.K., 2007. Bioherbicides: research and risks. Toxin Rev. 26, 313–342.

Hodgson, R.H., Wymore, L.A., Watson, A.K., Snyder, R.H., Collette, A., 1988. Efficacy of *Colletotrichum coccodes* and thidiazuron for velvetleaf (*Abutilon theophrasti*) control in soybean (*Glycine max*). Weed Technol. 2, 473–480.

Hoffmann, J.H., Moran, V.C., Impson, F.A.C., 1998. Promising results from the first biological control programme against a solanaceous weed (*Solanum elaeagnifolium*). Agric. Ecosyst. Environ. 70 (2–3), 145–150.

Hofmeister, F.M., Charudattan, R., 1987. Pseudocercospora nigricans, a pathogen of sicklepod (*Cassia obtusifolia*) with biocontrol potential. Plant Dis. 71, 44–46.

Holcomb, G.E., Jones, J.P., Wells, D.W., 1989. Blight of prostrate spurge and cultivated poinsettia caused by *Amphobotrys ricini*. Plant Dis. 73, 74–75.

Imaizumi, S., Nishino, T., Miyabe, K., Fujimori, T., Yamada, M., 1997. Biological control of annual bluegrass (*Poa annua* L.) with a Japanese isolate of *Xanthomonas campestris* pv. Poae (JT-P482). Biol. Control 8, 7–14.

Inman, R.E., 1971. A preliminary evaluation of rumex rust as a biological control agent for curly dock. Phytopathology 61 (1), 102–107.

Jackson, M.A., Bothast, R.J., 1990. Carbon concentration and carbon to nitrogen ratio influence submerged culture conidiation by the potential bioherbicide *Colletotrichum truncatum* NRRL 13757. Appl. Environ. Microbiol. 56, 3435–3438.

Javaid, A., Shafique, S., Bajwa, R., 2006. Effect of aqueous extracts of allelopathic crops on germination and growth of *Parthenium hysterophorus* L. S. Afr. J. Bot. 72 (4), 609–612.

Jayanth, K.P., 1987. Introduction and establishment of *Zygogramma bicolorata* on *Parthenium hysterophorus* at Bangalore, India. Curr. Sci. 56 (7), 310–311.

Jayanth, K.P., Bali, G., 1994. Biological control of *Parthenium hysterophorus* by the beetle *Zygogramma bicolorata* in India. FAO Plant Protect. Bull. 42 (4), 207–213.

Johnson, D.R., Wyse, D.L., Jones, K.J., 1996. Controlling weeds with phytopathogenic bacteria. Weed Technol. 10, 621–624.

Jones, L.A., Mandrak, N.E., Cudmore, B., 2017. Updated (2003–2015) biological synopsis of grass carp (*Ctenopharyngodon idella*). DFO Can. Sci. Advis. Sec. Res. Doc. 2016/102. Iv + 63 p.

Kazinczi, G., Lukacs, D., Takacs, A., Horvath, J., Gaborjanyi, R., Nadasy, M., Nadasy, E., 2006. Biological decline of *Solanum nigrum* due to virus infections. J. Plant Dis. Protect. Sonderheft XX, 325–330.

Kennedy, A.C., Elliott, L.F., Young, F.L., Douglas, C.L., 1991. Rhizobacteria suppressive to the weed downy brome. Soil Sci. Soc. Am. J. 55, 722–727.

Khan, M.A., Blackshaw, R.E., Marwat, K.B., 2009. Biology of milk thistle (*Silybum marianum*) and the management options for growers in north-western Pakistan. Weed Biol. Manage. 9, 99–105.

Klein, H., 2011. A catalogue of the insects, mites and pathogens that have been used or rejected, or are under consideration, for the biological control of invasive alien plants in South Africa. Afr. Entomol. 19 (2), 515–549.

Kohlschmid, E., Sauerborn, J., Müller-Stöver, D., 2009. Impact of *Fusarium oxysporum* on the holoparasitic weed *Phelipanche ramosa*: biocontrol efficacy under field-grown conditions. Weed Res. 49 (Suppl. S1), 56–65.

Lamberth, C., 2016. Naturally occurring amino acid derivatives with herbicidal, fungicidal or insecticidal activity. Amino Acids 48 (4), 929–940.

Lauriault, L.M., Kleeschulte, J., Michels, G.J., Thompson, D.C., 2013. First report of *Aceria malherbae* gall mites for control of field bindweed in Missouri. Southwest. Entomol. 28 (2), 353–356.

Madariaga, R.B., Scharen, A.L., 1985. Septoria Tritici blotch in Chilean wild oat. Plant Dis. 69, 126–127.

Marley, P.S., Kroschel, J., Elzien, A., 2005. Host specificity of *Fusarium oxysporum* Schlect (isolate PSM 197), a potential mycoherbicide for controlling *Striga* spp. in West Africa. Weed Res. 45 (6), 407–412.

Massion, C.L., Lindow, S.E., 1986. Effects of *Sphaacelotheca holci* infection on morphology and competiveness of johnsongrass (*Sorghum halepense*). Weed Sci. 34, 883–888.

McFadyen, R.E., McClay, A.R., 1981. Two new insects for the biological control of parthenium weed in Queensland. In: Wilson, B.J., Swarbrick, J.D. (Eds.), Sixth Australian Weeds Conference. Weed Science Society of Queensland, Australia, pp. 145–149.

Mortensen, K., Makowski, R.M.D., 1997. Effects of *Colletotrichum gloeosporioides* f. sp. malvae on plant development and biomass of non-target crops under field conditions. Weed Res. 37, 51–360.

Mortenson, K., 1998. Biological control of weeds using microorganisms. In: Boland, G.J., Kuykendal, L.D. (Eds.), Plant-Microbe Interaction and Biological Control. Marcel Dekker, New York, pp. 223–248.

Myers, J.H., Cory, J.S., 2017. Biological control agents: invasive species or valuable solutions? In: Impact of Biological Invasions on Ecosystem Services. Springer International Publishing, Switzerland, pp. 191–202.

Neumann, S., Boland, G.J., 1999. Influence of selected adjuvants on disease severity by *Phoma herbarum* on dandelion (*Taraxacum officinale*). Weed Technol. 13, 675–679.

Oehrens, E., 1977. Biological Control of Blackberry Through the Introduction of the Rust, *Phragmidium violaceum*. FAO Plant Protect. Bullet, Chile, p. 25. 26–28.

Olckers, T., 1995. Indigenous parasitoids inhibit the establishment of a gallforming moth imported for the biological control of *Solanum elaeagnifolium* Cav. (Solanaceae) in South Africa. Afr. Entomol. 3 (1), 85–87.

Olckers, T., Hulley, P.E., 1995. Importance of preintroduction surveys in the biological control of Solanum weeds in South Africa. Agric. Ecosyst. Environ. 52 (2–3), 179–185.

Olckers, T., Zimmermann, H.G., 1991. Biological control of silverleaf nightshade, *Solanum elaegnifolium*, and bugweed, *Solanum mauritianum*, (Solanaceae) in South Africa. Agric. Ecosyst. Environ. 37 (1–3), 137–155.

Pacanoski, Z., 2015. Bioherbicides. In: Price, A., Kelton, J., Sarunaite, L. (Eds.), Herbicides, Physiology of Action and Safety. InTechOpen, Rijeka, Croatia, pp. 253–274.

Pedras, M., Ismail, N., Quail, J., Boyetchko, S., 2003. Structure, chemistry, and biological activity of pseudophomins A and B, new cyclic lipodepsipeptides isolated from the biocontrol bacterium *Pseudomonas fluorescens*. Phytochemistry 62 (7), 1105–1114.

Phatak, S.C., 1992. In: Development and commercialization of rust (*Puccinia canaliculata*) for biological control of yellow nutsedge (*Cyperus esculentus* L.). Proceedings of the First International Weed Control Congress. Weed Society of Victoria, Melbourne.

Quail, J.W., Ismail, N., Pedras, M.S.C., Boyetchko, S.M., 2002. Pseudophomins A and B, a class of cyclic lipodepsipeptides isolated from a *Pseudomonas* species. Acta Crystallogr. Sect. C: Cryst. Struct. Commun. 58, 268–271.

Rana, S.S., Rana, M.C., 2016. Principles and Practices of Weed Management. Department of Agronomy. College of Agriculture, CSK Himachal Pradesh Krishi Vishvavidyalaya, Palampur, p. 138.

Riddle, G.E., Burpee, L.L., Boland, G.J., 1991. Virulence of *Sclerotinia sclerotiorum* and *S. minor* on dandelion (*Taraxacum officinale*). Weed Sci. 39, 109–118.

Ridings, W.H., 1986. Biological control of strangler vine in citrus: a researcher's view. Weed Sci. 34 (Suppl. 1), 31–32.

Rosenthal, S.S., Buckingham, G.R., 1982. Natural enemies of *Convolvulus arvensis* in Western Mediterranean Europe. Hilgardia 50 (2), 1–19.

Scheepens, P.C., 1987. Joint action of *Cochliobolus lunatus* and atrazine on *Echinochloa crus-galli* (L.) Beauv. Weed Res. 27, 43–47.

Scheepens, P.C., van Zon, H.C.J., 1982. Microbial herbicides. In: Kurstak, E. (Ed.), Microbial and Viral Pesticides. Marcel Dekker, New York, pp. 623–641.

Shabana, Y., 2005. The use of oil emulsions for improving the efficacy of *Alternaria eichhorniae* as a mycoherbicide for water hyacinth (*Eichhornia crassipes*). Biol. Control: Theory Appl. Pest Manage. 32, 78–89.

Shabbir, A., Dhileepan, K., Zalucki, M.P., O'Donnell, C., Khan, N., Hanif, Z., Adkins, S.W., 2015. The combined effect of biological control with plant competition on the management of parthenium weed (*Parthenium hysterophorus* L.). Pak. J. Bot. 47 (SI), 157–159.

Shamoun, S.F., Hintz, W.E., 1998. Development and registration of *Chondrostereum purpureum* as a mycoherbicide for hardwood weeds in conifer reforestation sites and utility rights of way. International Bioherbicide Workshop: Programme and Abstracts. University of Strathclyde, Glasgow, p. 14.

Sheppard, A.W., Shaw, R.H., Sforza, R., 2006. Top 20 environmental weeds for classical biological control in Europe: a review of opportunities, regulations and other barriers to adoption. Weed Res. 46 (2), 93–117.

Simmons, E.G., 1998. *Alternaria themesand* variations (224–225). Mycotaxon 68, 417–427.

Smith, J., Wherley, B., Reynolds, C., White, R., Senseman, S., Falk, S., 2015. Weed control spectrum and turf grass tolerance to bioherbicide *Phoma macrostoma*. Int. J. Pest Manage. 61, 91–98.

Smith, L., de Lillo, E., Amrine, J.W., 2010. Effectiveness of eriophyid mites for biological control of weedy plants and challenges for future research. Exp. Appl. Acarol. 51 (1–3), 115–149.

Smith Jr., R.J., 1991. Integration of biological control agents with chemical pesticides. In: TeBeest, D.O. (Ed.), Microbial Control of Weeds. Chapman and Hall, New York, pp. 189–208.

Stirk, W.A., Thomson, S.V., Staden, J.V., 2006. Effect of rust-causing pathogen (*Puccinia thlaspeos*) on auxin-like and cytokinin-activity in dyer's woad (*Isatis tinctoria*). Weed Sci. 54, 815–820.

Tanveer, A., Khaliq, A., Ali, H.H., Mahajan, G., Chauhan, B.S., 2015. Interference and management of parthenium: the world's most important invasive weed. Crop. Prot. 68, 49–59.

TeBeest, D.O., 1990. Conflicts and strategies for future development of mycoherbicides. In: Baker, R.R., Dunn, P.E. (Eds.), New Directions in Biological Control: Alternatives for Suppressing Agricultural Pests and Disease. New York, Alan R. Liss, Inc., pp. 323–332.

TeBeest, D.O., Yang, X.B., Cisar, C.R., 1992. The status of biological-control of weeds with fungal pathogens. Annu. Rev. Phytopathol. 30, 637–657.

Templeton, G.E., 1992. Use of *Colletotrichum* strains as mycoherbicides. In: Bailey, J.A., Jeger, M.J. (Eds.), Colletotrichum: Biology, Pathology and Control. CAB International, Wallingford, UK, pp. 358–380.

Terblanche, C., Nanni, I., Kaplan, H., Strathie, L.W., McConnachie, A.J., Goodall, J., Van Wilgen, B.W., 2016. An approach to the development of a national strategy for controlling invasive alien plant species: the case of *Parthenium hysterophorus* in South Africa. Bothalia-African Biodivers. Conserv. 46 (1), 1–11.

Tóth, P., Tóthova, M., Cagáň, L., 2008. In: Potential biological control agents of field bindweed, common teasel and field dodder from Slovakia. Proceedings of the XII International Symposium on Biological Control of Weeds. CABI, pp. 216–220.

Uludag, A., Gbehounou, G., Kashefi, J., Bouhache, M., Bon, M.C., Bell, C., Lagopodi, A.L., 2016. Review of the current situation for *Solanum elaeagnifolium* in the Mediterranean Basin. EPPO Bull. 46 (1), 139–147.

Van Lenteren, J.C., 2012. IOBC Internet Book of Biological Control, Version 6. International Organization for Biological Control, Wageningen, The Netherlands, p. 182.

Van Wilgen, B.W., Moran, V.C., Hoffmann, J.H., 2013. Some perspectives on the risks and benefits of biological control of invasive alien plants in the management of natural ecosystems. Environ. Manage. 52 (3), 531–540.

Vey, A., Hoagland, R.E., Butt, T.M., 2001. Toxic metabolites of fungal biocontrol agents. In: Butt, T.M., Jackson, C., Magan, N. (Eds.), Fungi as Biocontrol Agents: Progress, Problems, and Potential. CABI Press, Wallingford, UK, pp. 311–346.

Walker, H.L., 1981. *Fusarium lateritium*: A pathogen of spurred anoda (*Anoda cristata*), prickly sida (*Sida spinosa*) and velvetleaf (*Abutilon theophrasti*). Weed Sci. 29, 629–631.

Walker, H.L., 1983. Control of sicklepod, showy crotalaria, and coffee senna with a fungal pathogen. U.S. Patent 4, 390.

Walker, H.L., Tilley, A.M., 1997. Evaluation of an isolate of *Myrothecium verrucaria* from sicklepod (*Senna obtusifolia*) as a potential mycoherbicide agent. Biol. Control 10, 104–112.

Weyl, P.S.R., Martin, G.D., 2016. Have grass carp driven declines in macrophyte occurrence and diversity in the Vaal River, South Africa? Afr. J. Aquat. Sci. 41 (2), 241–245.

Whitney, N.G., Taber, R.A., 1986. First report of *Amphpbotrys ricini* infecting *Caperonia palustris* in the United States. Plant Dis. 70, 892.

Winder, R.S., Van Dyke, G.C., 1989. The pathogenicity, virulence, and biocontrol potential of two *Bipolaris* species on johnsongrass (*Sorghum halepense*). Weed Sci. 38, 84–89.

Winston, R.L., Schwarzländer, M., Hinz, H.L., Day, M.D., Cock, M.J.W., Julien, M.H., 2014. Biological Control of Weeds: A World Catalogue of Agents and Their Target Weeds, fifth ed. USDA Forest Service, Forest Health Technology Enterprise team, Morgantown, WV. FHTET-2014-04, 838 pp.

WSSA, 2017. WSSA Position Statement on Biological Control of Weeds, http://wssa.Net/wssa/weed/biological-control/ (accessed 14.03.17).

Zhang, W., Watson, A.K., 1997. Host range of *Exserohilum monoceras*, a potential bioherbicide for the control of *Echinochloa* species. Can. J. Bot. 75, 685–692.

Further Reading

CBD, 1992. Convention on Bioligical Diversity. CBD, United Nations (28 pp.).

Duke, S.O., Dayan, F.E., 2015. In: Discovery of new herbicide modes of action with natural phytotoxins. Am Chem Soc Symp Ser. vol. 1204, pp. 79–92.

Feuerherdt, L., Biosecurity, N.R.M., 2010. Management Plan for Silverleaf Nightshade (*Solanum elaeagnifolium*) in South Australia. Government of South Australia, Biosecurity SA.

Lu, X., Siemann, E., He, M., Wei, H., Shao, X., Ding, J., 2015. Climate warming increases biological control agent impact on a non-target species. Ecol. Lett. 18 (1), 48–56.

Morris, M.J., Wood, A.R., den Breeyen, A., 1999. Plant pathogens and biological control of weeds in South Africa: a review of projects and progress during the last decade. African Entomol. Memoirs 1, 129–137.

Seastedt, T.R., 2015. Biological control of invasive plant species: a reassessment for the Anthropocene. New Phytol. 205 (2), 490–502.

Simberloff, D., 2012. Risks of biological control for conservation purposes. BioControl 57 (2), 263–276.

8

Mechanical Weed Control

Mubshar Hussain*, Shahid Farooq[†], Charles Merfield[‡], Khawar Jabran[§]

[*]Bahauddin Zakariya University Multan, Multan, Pakistan [†]Faculty of Agriculture, Gaziosmanpaşa University, Tokat, Turkey [‡]The BHU Future Farming Centre, Canterbury, New Zealand [§]Düzce University, Düzce, Turkey

8.1 INTRODUCTION, BRIEF HISTORY, AND MECHANISM

Agriculture is the foundation of civilization and has made unlimited contributions to all human societies. Domestication of plant species has benefited human beings and helped them achieve their basic food and animal feed needs (Kareiva et al., 2007; Meyer et al., 2012; Zohary et al., 2012). It was however coupled with the infestation of agricultural fields with unwanted plant species, which are now termed as weeds (De Wet and Harlan, 1975). In this way, weeds have been an inescapable part of agricultural ecosystems along with the crop plants since the start of agriculture (De Wet and Harlan, 1975; Zohary et al., 2012). Mechanical weed control through tillage has been the major weed management method for many centuries (Timmons, 1970). Although tillage provides numerous benefits, weed control may be considered as its major objective. Weeds were probably assessed as the undesired plants by the earlier people, and hence, they used hand pulling as the pioneer weed control method (Timmons, 1970). This probably had happened 8000–10,000 years ago. The methods following hand weeding included the use of knives or similar tools for cutting weeds, the use of hoes for weeding, and the use of plows or hoes for weeding, all being the forms of mechanical weed control (Hamill et al., 2004). These hoes and plows were modified to animal-driven implements as area under cultivation was increased. Later on, in human history, improvement in mechanical weed control has been mostly linked to industrial advancement. An important aspect is the involvement of agricultural labor in the industry with the industrial development that emphasized the need for weed control methods requiring low labor inputs. Animal-drawn cultivation tools had reduced the labor requirements (compared with hand weeding) and reduced the time required for soil manipulations and weed control. The horse-drawn cultivator made by Jethro Tull (1674–1741) provided the basis for revolutionizing mechanical weed control. Starting from England, the work of Tull helped to improve mechanical weed control in all parts of the world.

Timmons (2005) reviewed the history of weed control in Canada and the United States. Weeds had been generally ignored as a pest problem. Nevertheless, tillage aimed at seedbed preparation also helped to suppress weeds in crop fields. Globally, mechanical control had been a major method for suppressing weeds before the 20th century, while Weed Science witnessed a major progress during the 20th century (Appleby, 2005; Timmons, 2005). Basic forms of many modern cultivation tools were initially developed for animal draft, and with the development of tractors, these were modified as tractor-driven machines. This greatly eased the process of soil cultivation aimed at weed control and seedbed preparation. The precision of weed control was increased after the development of tools that could provide cultivation in a close vicinity (e.g., ~5 cm) of plant rows in a row crop. There is little information available regarding the development of weeding tools in different parts of the world. It is considered that most post-1960 weeding tools have been concurrently developed in Europe and Asia (with there being limited exchange of practical information or weeding machinery between Asia and the EU), with a few novel tools developed in North America.

Organic farming over the past 40 years has stimulated the development of mechanical weeding methods (Pannacci et al., 2017; Pannacci and Tei, 2014). Mechanical weeding is now effective and fast, and leaves no chemical/herbicide residues on crop plants or soil. Modernized mechanical weed control, particularly when using automated guidance systems, can be a highly effective substitute for chemical weed control. Precise mechanical weed control has been suggested by researchers in the recent past. This allows mechanical control close to plants (through vision guidance and sensors discriminating between weeds and crop plants) in the crop rows (Wiltshire et al., 2003). Precise mechanical hoeing provided weed control and grain yield comparable with chemical weed control in sugar beet (Wiltshire et al., 2003). Today's vision-guided weeders have the potential to revolutionize weed control. These tools can be divided into discriminatory weeders that use sensors (computer vision, light detection, and ranging sensors) to discriminate weeds from crops and only apply the weeding tool to the weeds and nondiscriminatory weeders such as torsion weeders and finger weeders where the weeding tool is applied to both weeds and crop, with the crop surviving due to a level of "resistance" to the tool. In intelligent weed control, weeds may be identified through the use of spectral reflectance and laser sensors based on the morphological characters of weeds.

8.2 MECHANISMS OF MECHANICAL WEED CONTROL

According to Chicouene (2007), mechanical weed control damages weeds by three ways: (1) cutting, (2) uprooting, and (3) burial. Mechanical weed control can cause cuts to any of a weed tissues (particularly in the case of young seedlings) that leads to exhaustion of plant reserves and hence the desiccation and withering of plants (Chicouene, 2007). In the process of uprooting, either the roots are exposed to the atmosphere, or the plants get completely detached from the soil; both lead to the withering and death of the plants (Chicouene, 2007). Burying is another way of mechanical damage to weed plants. For instance, in the rice farming, the soil is puddled before the sowing of the crop that helps to bury majority of the emerged or emerging weeds. Mechanical weed control has the potential to supplement other weed control methods. Similarly, it can also be integrated with other methods to achieve integrated weed control.

8.3 MERITS AND ENVIRONMENTAL ISSUES

Mechanical weed control is an effective weed management method in row crops, organic farming, and minor crops (like vegetables, fruits, or some seed crops). Mechanical weeding can provide effective weed management even when other methods are not possible and can outperform them in some situations. There are several forms of mechanical weed control as described earlier ranging from handheld tools to the most advanced vision-guided hoes. Hand hoeing may be used to control weeds in small-scale farming as it is less safe to use herbicides in home gardening due to the lack of training. For example, insufficient time between herbicide application and harvest and consumption of vegetables may be a health hazard or where the application of a wrong herbicide may cause killing of the crop. Hence, in home gardening, hand hoeing is the preferred weeding technique and achieves complete weed control.

While there are many positive aspects of mechanical weed control, there are negative effects as well. These include (i) cost and time required that can impact other crop operations, (ii) effectiveness that is highly dependent on weather and soil conditions and correct time of application, (iii) lower efficacy of intrarow weed control, (iv) required skilled labor, and (v) high capital cost. There is a general assumption that weeds growing in intrarows are left uncontrolled by mechanical weeding. However, recent advances have resulted in the development of several implements (discriminatory and nondiscriminatory) such as finger weeders, brush weeders, torsion weeders, mini ridgers, and computer-vision-guided hoes that provide excellent control of intrarow weeds. Mechanical weed control can also damage plant roots. In fruit orchards, the trees' feeder roots are present in the soil surface to absorb nutrients. Tillage aimed controlling the weeds in fruit orchards can damage these roots. Trunks of trees can also be damaged by the mechanical weed control implements if they have been adjusted incorrectly. Moreover, the operation of mechanical weed control may face limitations owing to adverse weather conditions. Tillage aimed at seedbed preparation provides weed control by burying weed seeds deep in the soil that become unable to germinate and mediate seed dormancy, longevity, and seedling emergence. Secondly, tillage leads to the germination of weed seeds that are near to the soil surface. The emerged weeds 2–3 weeks after the crop sowing can be successfully controlled by mechanical harrowing. Several demerits are also associated with tillage (such as soil disturbance, high cost, and fuel consumption); therefore, conservation agriculture has been devised as an alternative technique to manage weeds and improve crop yields (Farooq et al., 2011). Conservation agriculture uses zero tillage or minimum tillage that results the emergence of weed seeds only near to soil surface (Singh et al., 2015). Minimum or zero tillage can consistently decrease the soil seed bank if emerging weeds after crop sowing are properly controlled. Different implements used in mechanical weed control, their working principles, and efficacy in different crops are described below.

8.4 HOES

Manual weed control through using different handheld implements/tools is the choice of low-income, small-land-holding farmers around the globe. Manual weed control, when carried out properly, is very effective, however extremely labor-intensive. Manual weed control through hand tools can only be employed in small-scale farming or in home gardens. Hoes are ancient and versatile hand tools used to shape soil, control weeds, and harvest the root crops since the start of agriculture. The hoes aiming at weed control either agitate the soil or cut the foliage of weeds, both leading to mortality of weed plants. Different kinds of hoes have varied shapes and purposes. Some of the hoes are multipurpose, while others are designed for specific/unique functions. The hoes are broadly classified into two generations, draw hoes—used for shaping the soil—and scuffle hoes, used for weed control. As the chapter focuses on mechanical weed control, therefore, it focuses on scuffle hoes. Scuffle hoes mainly have two designs, the Dutch hoe and the Hoop hoe.

8.4.1 The Dutch Hoe

The Dutch hoe is designed to be pushed or pulled through the soil to cut the roots of weeds just under the surface. It has a sharp blade (on both sides) to cut on both forward and backward strokes (Loudon, 1871). The blade is pushed within the soil at a shallow depth to cut the roots of weeds. The hoe can be used for dual purposes, that is, to cultivate the soil and remove weeds. The angle of the hoe can be adjusted to push the hoe at shallow depth.

8.4.2 The Hoop Hoe

The Hoop hoe is also known as the oscillating hula, stirrup, pendulum weeder, or swivel hoe (USAID, 1984). It has a double-edge blade attached to a shaft and blades bend around to form a rectangle. The push or pull movement of the blades controls weeds just below the soil surface. The blades are usually 7–15 cm wide.

These hoes can be used in agricultural areas (small scale) and home gardening. There are some specific types of implements that are used for removing woody weeds especially from nonagricultural areas.

8.4.3 Spades and Hand-Held Hoes

Several handheld spades and hoes are used to eradicate weeds from row crops or vegetables grown at small scale. These tools are used for controlling weeds growing between crop plants (Ampong-Nyarko and De Datta, 1991; Tu et al., 2001). The level of mechanical weed control largely varies between developing and developed countries. In the developed world, there are advanced implements that are efficient in intrarow weed control. However, the absence of these implements in developing world makes intrarow weed control difficult, and farmers depend on handheld hoes and spades. The handheld tools such as hoe, narrow spade, Swiss hoe, knife, machete, and pointed stick are used in many countries (in the developing world) to remove the weeds growing within crop rows. However, these tools need

FIG. 8.1 A typical hand hoe (A) used to eradicate weeds from crop rows and (B) working of the hoe in standing crop. *Photograph, Mubshar Hussain.*

intensive labor work, and weed removal from 1 ha could take 10–30 days of labor input (Ampong-Nyarko and De Datta, 1991). The handheld hoe and its working have been shown in Fig. 8.1. The selection of handheld tools for weed removal depends upon the nature of crop grown, row-to-row distance, and intensity of weed infestation. Due to intensive labor requirements, these tools are not feasible for commercial farming; rather, small-scale farmers and kitchen gardeners use these implements to remove weeds manually. On the other hand, the availability of advanced mechanical weeding implements in the developed world omits the use of handheld hoes and spades for mechanical weed control.

8.4.4 Root Talon

Root talon is an inexpensive and lightweight hand tool used to eradicate shallow-rooted weeds (Tu et al., 2001). It is shaped like a pickax with a plastic handle and metal head (Fig. 8.2). The specialized claw and gripping device allows the users to pull up and remove weeds. Root talon has successfully been used to eradicate weeds like *Ailanthus* spp., *Sesbania punicea* (Cav.) Benth., and *Rhamnus* spp. Root talon is not effective against deep-rooted and spiny weeds (Tu et al., 2001).

8.4.5 Weed Wrench

Weed wrenches are made of steel instead of the plastic handles of the root talon and are designed to grip the stem rather than cut into the soil as the root talon does. The weed wrench consists of a long handle and a jaw (Fig. 8.3). The stem of the weed is placed in the jaw, and the handle is used to pull the weeds up. The steel frame is capable of withstanding more strain than root talon and can be used to pull up deep-rooted weeds. Weed wrenches are available in

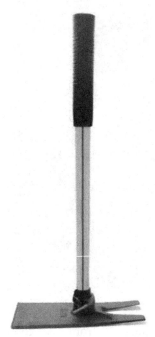

FIG. 8.2 Root talon used to pull up the shallow-rooted weeds. *Source: http://kickapoowoods.org/member-services/ forestry-store/.*

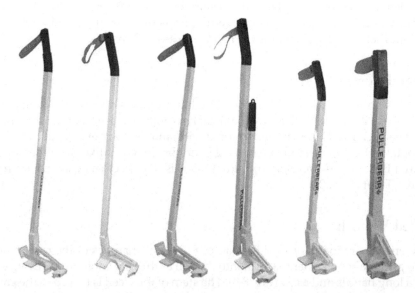

FIG. 8.3 Weed wrenches used to eradicate deep-rooted weeds. *Source: https://homemadewilderness.com/category/ invasive-species/.*

different sizes, and size selection depends upon the target weed (Tu et al., 2001). Weed wrenches can pull up any plant as long as the stem of the particular plant fits into the jaws of weed wrench and have more strength than anchoring roots, and stem is not damaged by the pressure of jaws. Weed wrenches have been successfully used to pull *Acacia melanoxylon* R. Br., *Rhamnus cathartica* L., *Elaeagnus angustifolia* L., *Rosa multiflora* Thunb., *Salix* spp., *Tamarix* spp., *Lonicera* spp., *Cytisus scoparius* (L.) Link, *Genista monspessulana* (L.) L.A.S. John-son, and *Schinus terebinthifolius* Steud. (Tu et al., 2001).

8.5 HARROWS AND TRACTOR HOES

A wide range of harrows and tractor-mounted hoes are used to control weeds in different crops. These range from simple harrows to the most advanced camera and global positioning system (GPS)-equipped robotic weeders. The available mechanical weeders can broadly be classified into two categories, that is, contiguous weeders that weed the whole field surface, both interrow and intrarow, and incontiguous weeders that apply different weeding tools to the interrow and intrarow. Contiguous weeders include CombCut, rotating-wire weeders, spoon weeders (rotary hoes), and spring tine harrows, while incontiguous weeders include the standard interrow hoe using horizontal metal blades, brush weeders, and basket weeders, weeding the interrow, and finger weeders, torsion weeders, mini ridgers, wire weeders, and pneumatic weeders, weeding the intrarow. Intrarow weeders are further classified into nondiscriminatory (low technology) and discriminatory (high technology). The nondiscri-minatory weeders control the intrarow weeds by applying the weeding tool to crop and weeds alike, so the crop has to be tougher ("resistant") than the weeds to survive, whereas discriminatory weeders discriminate crop from weeds using systems such as computer vision or sensor wands and then selectively apply the weeding tool only to the weeds. A number of nondiscriminatory weeders are available for intrarow weeding in different crops (van der Schans et al., 2006). Recently, cultivators equipped with sensor and mapping techniques—so-called intelligent cultivators or discriminatory weeders—have become a viable option for cultivation between crop rows and ultimately weed control (Slaughter et al., 2008; Christensen et al., 2009; Griepentrog et al., 2006). Adequate literature is available regarding the efficacy of low-technology mechanical weeders (Rasmussen, 2004; Rasmussen and Ras-mussen, 2000; Melander et al., 2015; Weber et al., 2016; Rueda-Ayala et al., 2015; Rasmussen et al., 2012; Van der Weide et al., 2008); however, literature on the efficacy of intelligent cul-tivators is limited (Tillett et al., 2008; Kunz et al., 2016; Gerhards et al., 2016).

The available interrow cultivation machines (Figs. 8.4 and 8.5) are capable of controlling 80% of the crop weeds (Nørremark and Griepentrog, 2004), while the remaining 20% of weeds grow and remain intact in intrarows (Figs. 8.6 and 8.7). Those weeds growing close to crop plants incur the heaviest losses on crop yields and are the most challenging for farmers. Dif-ferent intrarow weeders including rolling harrows, basket weeders, rolling cultivators, and rotary cultivator are viable options to manage the intrarow weeds (Cloutier et al., 2007). Some interrow weeders can achieve a level of intrarow weed control, by moving soil into the intrarow. Rolling cultivators can be used for intrarow weed control by moving soil into the crop row, but the amount of soil they move is difficult to accurately control, so they

FIG. 8.4 Steerage hoe (A) used to control interrow weeds in row crops and its working in the field (B). *Photograph, Charles N Merfield.*

FIG. 8.5 Interrow cultivator used for physical weeding (A) its component (B) and crop row adjustment (C) to avoid the damage to crop plants. *Photograph, Charles N Merfield.*

FIG. 8.6 Moving-wire weeder (A) used for intrarow weeding and its component (B). *Photograph, Charles N Merfield.*

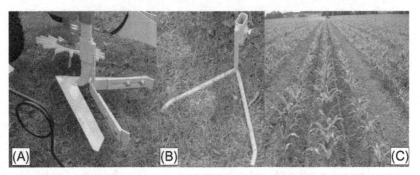

FIG. 8.7 Mini ridger (A) equipped with components of finger and torsion weeder (B) and weed-free maize crop following intrarow weeding with mini ridger (C). *Photograph, Charles N Merfield.*

can only be used in more robust crops, for example, maize and potatoes, or at larger sizes of crops, for example, cabbages.

However, the intrarow weeds are difficult to control without injuring crop plants (Tillett et al., 2008). Finger and torsion weeders are the commonly available mechanical weeders to control weeds in the intrarow. However, for three-point linkage mounted machines, a second person, in addition to the tractor driver, is required to steer these implements. However, if a tool carrier or automatic guidance system is used, only the tractor driver is required. Moreover, required level of placement accuracy varies among the tools; torsion weeders need particularly accurate placement, while finger weeders require relatively less accuracy of placement. A vertical axis brush weeder was developed for intrarow weeding that was developed but failed commercially due to its cost (due to multiple hydraulic motors); it was difficult to get a consistent result brushing into the rows, and other tools, such as mini ridgers, could do the same job cheaper and more reliably. Advanced sensing technologies have been applied to identify the crop plants and weeds and then weed species. These sensors then identify the weeds of particular interest based on their morphology and eradicate the weeds (Young et al., 2014).

8.5.1 Harrowing

Weed harrowing, a mechanical weed control method, is being used for weed control since the 20th century (Korsmo, 1926). Harrowing is a mechanical cultivation method applied to both crop and weed plants. Harrows are contiguous weeders working in both the intrarow and interrow. Weed harrows have undergone intensive modernization with respect to tine depth and angle; however, their applicability to early crop stages is still limited (Van der Weide et al., 2008). Modern spring tine harrows were designed for weeding dicot weeds in monocot crops, that is, cereals. In cereals, their use at early crop stages is generally not problematic if done carefully. When used in dicot arable crops, for example, peas, beans, and linseed, much greater care is required at early crop stages and often all crop stages. Spring tine harrows can also be used in some vegetable crops. Working principle and selectivity of harrows have been described in detail by Kurstjens and Kropff (2001), Kurstjens (2002), and Cirujeda et al. (2003). Typical mechanical harrows and their components used in different

FIG. 8.8   Mechanical harrow in action (A) and its components (B and C). *Photograph, Charles N Merfield.*

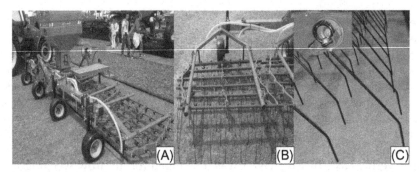

FIG. 8.9   Mechanical harrow manufactured by Hatzenbichler (A), its working in the field (B), and component (C). *Photograph, Charles N Merfield.*

FIG. 8.10   Mechanical harrow manufactured by Einbock (A) for use on commercial farms, its working in the field (B), and folded view (C). *Photograph, Charles N Merfield.*

regions of the world are shown in Figs. 8.8–8.10. Harrowing can be practiced either as pre-emergence or during postemergence stages of crop plants. Preemergence harrowing is gentle and applied to deeply sown crops. Postemergence harrowing is capable of controlling small weeds, which have not passed their first true-leaf stage (Van der Weide et al., 2008).

Early emerging annual weeds are either uprooted or covered by preemergence mechanical harrowing (Kurstjens and Kropff, 2001). Preemergence harrowing kills the weeds from white thread to the cotyledon stage. However, operating speed and tines can be adjusted to work the harrow more vigorously and kill bigger weeds in deep-rooted crops such as beans, peas, sweet maize, and spinach (Van der Weide et al., 2008). The adjustment of operating speed and tine could control the small-seeded broad-leaved weeds having 2–4 true leaves up to 90% (Van der Weide and Kurstjens, 1996). For postemergence harrowing, spring tine harrows are used in different crops such as cereals (Rasmussen and Ascard, 1995), maize (Baumann, 1992), potatoes (Rasmussen, 2002), peas, beans and many transplanted vegetables, sugar beet (Hallefalt et al., 1998), and carrots (Fogelberg, 2007). Harrowing cannot be practiced in sensitive crops such as sugar beet before the four-leaved stage of the crop (Ascard and Bellinder, 1996; Westerdijk et al., 1997). Various initiatives have also been started in different countries to promote harrowing in the rising trend of herbicide use and associated risks. For example, farmers are subsidized for using harrowing in parts of Norway (Fylkesmannen, 2011).

Preemergence harrowing is less damaging to crop plants compared with postemergence harrowing as at preemergence stage crop plants are not yet emerged (Lundkvist, 2009). Brandsæter et al. (2012) also concluded that preemergence harrowing increased the average crop yield by 6.2% while improved by 4.0% after postemergence harrowing. Similarly, Armengot et al. (2013) concluded that harrowing prevents emerging weeds from being a limiting factor for crop productivity in organic cereal fields.

In mechanical weeding, the critical period to control weeds is early in the crop's growth because it is the weed size that is important for timing of weed control compared with herbicide-based weed management where crop size more often determines timing of weeding. Generally, the optimum weed size for mechanical weeding is cotyledon stage; as before, these (the white-thread stage) weeds are better able to survive burial, and after (true-leaf stages), the plant is bigger and physically more resistant. It is therefore concluded that weed harrowing is a viable tool to manage weeds at earlier stages of crop growth when weed size is small.

8.5.2 Brush Weeders

Brush weeders use flexible brushes made up of nylon rotating around a horizontal axle. The brushes are rotated using power take-off from the tractor. Brush weeders are very aggressive—they cause the uprooting of weeds (through pulverize soil), burying, breaking, and destroying the weeds. The whole machine and working parts of brush weeder are shown in Fig. 8.11. A protective shield or cover is normally installed to keep the crop from being damaged. These weeders, like other mechanical weeders, also require accurate steering to minimize the damage to crop plants (Melander, 1997; Cloutier et al., 2007).

8.5.3 Finger Weeder

The finger weeder is a simple, nondiscriminatory intrarow mechanical weeder. The original design produced by Bezzerides in the United States has rubber "fingers" mounted on

FIG. 8.11   Brush weeder (A) and its working in different row crops (B). *Photograph, Charles N Merfield.*

wires attached to pairs of steel cones with soil-engaging spikes to turn them. The sandwich design (developed by Kress & Co and Wageningen University) consists of two plates, a steel one with bent fingers that engage the soil and drives it round and a second flat one that makes up the weeding fingers. The fingers are made from a range of materials, including steel, hard and soft plastics, fabric-reinforced rubber, and even brushes. A third approach is to create the entire weeder from a single piece of injection-molded plastic. Fingers are horizontally pointed outward, and ground-driven rotary motion enables them to function from sides and beneath the crop plants. A typical finger weeder is shown in Fig. 8.12. Fingers remove the weeds by penetrating just beneath the soil surface with minimum damage to the crop plants if set up correctly. However, most intrarow weeders will cause crop damage if incorrectly set up or used. Fingers perform best in loose soil, while the performance is poor in heavy soils and crusty surface after rainfall. Finger weeders are especially successful against weeds at cotyledon stage (Cloutier et al., 2007). All of mechanical weeders give the highest level of weed kill at cotyledon stage of weeds, as that is the point when the plant has used up seed reserves and is at its smallest and weakest. Finger weeders can successfully be used in many transplanted vegetables, beans, spring-seeded rape, seeded onions (two-leaf stage and beyond), red beet

FIG. 8.12   Finger weeder (A) and its component—hard plastic finger (B). *Photograph, Charles N Merfield.*

and sugar beet (two-to-four-leaf stage), carrots (two-leaf stage and beyond), and strawberries (van der Schans et al., 2006; Cloutier et al., 2007).

Finger weeder has been found to control the intrarow weeds in organic corn with 61% efficacy (Alexandrou, 2004). The core requirement of the finger weeder is that tractor should be steered in the right position, so that fingers can effectively operate to control the target weeds (Bowman, 1997; Cloutier et al., 2007; Van der Weide et al., 2008). All mechanical weeders require accurate steering, and weed control depends upon the level of steering. Finger weeder can perform well with middle level of steering. Cirujeda et al. (2015) evaluated the performance of finger weeder for 7 years in processing tomato and concluded that the weeder had irregular performance due to soil crust and additional implements or repeated operations that were needed for successful weed management in the crop. Indeed, all mechanical weeders are required to be used in combination with other weed control methods for ultimate weed control. Similarly, Kunz et al. (2016) tested the efficacy of finger weeder to suppress weeds (both inter and intrarow weeds) compared with common hoeing (manual steering of Einböck CHOPSTAR, Dorf an der Pram, Austria hoe) and found that finger weeder (with automatic steering) suppressed 29% more weeds in sugar beet crop. However, there were no significant differences for suppression of intrarow weeds among manual steering and finger weeders.

8.5.4 Torsion Weeder

Torsion weeder is another mechanical weeder used for intrarow weeding. The torsion weeder has a rigid frame having spring tines that are connected and bent (Fig. 8.13). This arrangement ensures that short tine segments are parallel to soil surface and meet near the plant row. In this way, the crop plants get passed through the tines without facing any damage if accurately steered. Torsion weeders are considered difficult to set up due to low steering tolerances, that is, if the ends of the torsion wires are slightly out of alignment, they can fail to kill any weeds and/or then can kill crop plants. The torsion weeder and its components are shown in Fig. 8.12. The coiled spring tines allow the tips to flex with soil contours and around established crops. A reduction of 60%–80% in weed densities has been observed by the use of torsion weeders (Cloutier et al., 2007). Torsion weeders, unlike finger weeders, require accurate, high-level steering to avoid damage to crop plants. Torsion weeders are often used with

FIG. 8.13 Torsion weeder (A) and its component (B). *Photograph, Charles N Merfield.*

precision, vision guidance systems using both computer guidance and real-time kinematic (RTK) GPS for more effective weed management (Bowman, 1997; Cloutier et al., 2007; Van der Weide et al., 2008). Torsion weeders have been tested in Europe and North America on many herbaceous perennial crops, and excellent results were observed in terms of intrarow weed control (Cloutier et al., 2007). Some reports also indicate that torsion weeders can also be successfully used in poor-rooted crops such as carrot and reduce weed density in the rows by 60%–80% (Peruzzi et al., 2005).

8.5.5 Mini-Ridgers

A third type of intrarow weeder is the "mini ridger" (Merfield, 2014, Fig. 8.14). Unlike most other weeders that predominately kill weeds by cutting and/or uprooting, mini ridgers exclusively kill weeds through burial. This is because once a weed seedling has emerged and opened its cotyledon leaves, it loses the ability it had as a germinating seed to grow up through soil. This means the ability of mini ridgers to kill weeds is unaffected by weather, while most other tools suffer decreased effectiveness in cool and wet conditions. As the ridge is created along the entire intrarow, mini ridgers can also kill close to crop plant weeds that discriminatory weeders cannot. The limitation of mini ridgers is that it works best on small weeds, from cotyledon stage to one to two true leaves, and that the crop plants must be larger than the ridge; otherwise, they will also be killed if they are buried. However, by using blades of different heights, from 10 mm to over 100 mm, the size of the ridge can be precisely controlled so allowing ridges to be placed under crop plants as small as 50 mm high while also creating large ridges at larger crop sizes.

8.5.6 Automation in Intrarow Weeding

Automation of intrarow weeding has greatly improved the efficacy and accuracy in mechanical weed control. Automation requires both detection (and differentiation of weeds and crop plants) and removal of weeds. The automated mechanical weeders are able to differentiate between crops and weed plants and remove weeds with precisely controlled devices (Bakker, 2009). Automatic mechanical weeders employ four different technologies for precision weed control, that is, (i) guidance, (ii) detection and identification, (iii) precision, and (iv) mapping (Slaughter et al., 2008). Mechanical knives traveling in and out of the crop rows or height adjustable rotating hoes are used in mechanically automated weed control (Astrand and Baerveldt, 2002). A vision-guided hoe is shown in Fig. 8.15.

FIG. 8.14 Mini ridger in cabbages (A) and different height blades (B) attached to "telescope" depth control system (C). *Photograph, Charles N Merfield.*

FIG. 8.15 Different views of Robocrop, a vision-guided machine for weed control (A and B). *Photograph, Charles N Merfield.*

The automation in mechanical weed control has improved weed control, and a number of automated mechanical weeders, termed as intelligent weeders or high-technology weeders, have been developed in various parts of the world. These weeders use computer vision, 3-D imaging, image algorithms, GPS, and cameras for detection, differentiation, and identification of weeds and crop plants (Merfield, 2016). Weed control with intelligent weeders has also been suggested as an effective technique against herbicide-resistant weeds (Prince et al., 2012). It is an evolutionary step in weed control, and the future of weed control with intelligent weeders depends on weed scientists and farming communities. A lot of investment has been made in the development of intelligent weeders, and several are now commercially available. The lack of development resources and subsequent commercialization are the major hurdles in sophisticated intelligent weed control. The lack of development problem (i.e., differentiating between weeds, crop, and soil) in vision-guided discriminatory weeders is another hurdle in intelligent weed control. However, with the limited resources, a few intelligent weeders have been commercialized that are discussed below.

A weeding machine using computer vision to sense plants was tested by Tillett et al. (2008). The machine is composed of a rotating half circle disk that avoids contacting crop plants during weeding process. The machine mounted a camera on the top at the height of 1.70 m, and field view was below the camera. Computer vision was used to detect the crop plants and determine the location of rotating disk with reference to crop plants. The machine was tested to control the intrarow weeds of cabbage plants at 16, 23, and 33 days after planting (DAP). The machine resulted in 77% and 87% reduction in the number of weed plants at 16 and 23 DAP, respectively. This machine has been commercialized under the name Robocrop (Inman, 2010).

A vision-based mobile intelligent weeder for perception and subsequent weed control was developed by Astrand and Baerveldt (2002). The machine used two cameras and a weeding tool, which was a rotating wheel oriented perpendicular to crop plants. One gray-scale camera located at the front of the machine was used to obtain the images of crop rows, while the second color camera obtained the pictures of crop plants. The camera successfully detected the crops using image segmentation techniques and differentiated weeds and crop plants. The efficacy of the machine was not reported.

An intrarow hoe developed by a French firm has been described by Cloutier et al. (2007). The light reflected from the field surface was sensed by this automated weeder to identify the crop plants. The weeder also used a system to control the hoe motion around crop plants. This weeder can effectively control weeds when they are smaller from crop plants, and efficacy is poor when the height of weeds exceeds from that of crop plants.

Gai (2016) used the fusion of two-dimensional textural images and three-dimensional spatial images to develop a weeder for recognizing and localizing crop plants at different growth stages. The 3-D photographs of weeds and crop plants were taken at different growth stages, and feature extraction algorithms were developed. These algorithms were tested to identify/differentiate crop plants from weeds in broccoli and lettuce crops. The weeder has 93.1% and 92.3% true positive detection rate for broccoli and lettuce, respectively. It was concluded that 3-D imaging-based plant recognition algorithms are effective and reliable for crop/weed differentiation and can be used for successful weed control.

Sabanci and Aydin (2017) developed an automatic weeding tool for weed control in sugar beet using image processing algorithms. The weeder was used for herbicide spraying with nozzles adjusted at different heights from the ground (30 and 50 cm). The herbicide efficacy was highly dependent on the speed of weeder and nozzle height. Photographs of the surface area covered by fluid sprayed by nozzles were taken and processed. It was concluded that surface area covered by the herbicides was inversely proportional to the speed of the weeder. The authors argued that the system can be mechanically developed for hoeing in sugar beet.

Although several examples of the automatic weeding tools for weed control are present in literature and are termed as "robotic weeders," however, the term "robotic weeding" has been challenged recently (Merfield, 2016). The author contends that the several or all commercially available so-called "robotic" weed controllers are actually not robots; rather, these are self-guiding vehicles carrying weeding tools. The term robotic weed control is a far more difficult objective. The author has argued that robotic weeders must be able to operate without human intervention and have 10 different qualities that start from monitoring the crop, weather, and soil and end with automatically returning, cleaning, and storing themselves in the farmyards (for details, see Merfield, 2016).

It is evident that a lot of advancements have been made in the intelligent weed control; still, more advancements are needed to effectively use them without damaging crop plants.

8.6 MOWING, CUTTING AND STRIMMING

Weeds exist in all human and natural environments, such as agriculture, parks and gardens, road corridors, and wildlife areas. The occurrence of weeds in areas of high conservation, urban areas, and parks needs weed management approaches that have low environmental impacts.

Weeds that are distributed on the field boundaries may spread and colonize the adjacent areas (Marshall, 1989; Cordeau et al., 2012) and therefore needed to be managed with low environmental impacts (Blackshaw et al., 2006; Chikowo et al., 2009). Mowing, cutting, and string trimmer are weed control methods with lower environmental impact. However, mowing can be a potential source of spread of seeds from one place to another. Weed mowing at seed set stage can potentially disperse seeds of the weeds to adjacent places, which can be a potential cause of seed spread. However, mowing has been concluded as an effective technique to reduce soil seed bank of an invasive weed (Milakovic and Karrer, 2016). The mowing technique has been effectively used in controlling invasive weeds along the roadsides in different parts of the world (Milakovic and Karrer, 2016; Jeffries, et al., 2016). Jeffries et al. (2016) evaluated the effects of mowing regimes and timing before herbicide application on *Paspalum urvillei* Steud. found on roadsides of North Carolina. The plant was able to survive under all tested treatments even 52 weeks after transplanting. The authors concluded that long-term mowing for many years is required to eradicate the plant. Similarly, Milakovic and Karrer (2016) evaluated the effect of different mowing regimes on the weed seed bank of six different *Ambrosia artemisiifolia* L. populations that were present on the roadsides of Austria. After 3 years of experimentation, it was concluded that mowing significantly reduced the weed seed bank from 45% to 80%.

The effects of mowing on aquatic weed species are particularly studied where one particular species may have blocked the water channels (Dawson, 1989, 1976; Pitlo and Dawson, 1990; Kern Hansen and Dawson, 1978), whereas the effect of mowing on weed communities is scarce and geographically restricted (Baattrup-Pedersen et al., 2002, 2003; Baattrup-Pedersen and Riis, 2004). Tarasoff et al. (2016) evaluated the effects of aggressive mowing on an invasive aquatic emergent plant, *Iris pseudacorus* L. at two locations within British Columbia, Canada (Vaseux Lake and Dutch Lake). Aggressive mowing was proved an effective method of controlling the invasive plant and it was concluded that aggressive mowing could effectively be utilized to control the plant.

8.7 PRACTICAL EXAMPLES FROM CROPS GROWN GLOBALLY

The different mechanical weed control techniques described above have successfully been used to control weeds in most arable and vegetable crops around the world. The methods of weed control in crops and the countries where these methods have been adapted are summarized in Table 8.1. Sugar beet, soybean, and maize are the crops where these methods have been used extensively (Table 8.1). Moreover, intensive work on mechanical weed control has been conducted in countries like Denmark, Germany, the United States, and the United Kingdom (Table 8.1). Different types of mechanical methods to manage the weeds of different crops are described below.

8.7.1 Sugar Beet

Kunz et al. (2015) evaluated the effects of herbicide application, mechanical hoeing, and precision steering on weed control and yield of sugar beet. Precision steering and mechanical hoeing not only reduced the herbicide input but also improved sugar beet yield up to 30%. In another study, Kunz et al. (2016) evaluated weed control in sugar beet through different mechanical weeders. Automatic steering technologies reduced the weed density by 82%. The use of finger weeders, rotary harrow, and torsion finger weeder with automatic steering reduced

TABLE 8.1 Different Mechanical Weed Control Methods Used in Different Crops Around the World

Country	Crop	Mechanical Weed Control Method	Reference
Denmark	Winter wheat	Weed harrowing, interrow weed hoeing	Rasmussen (2004)
Denmark	Barley	Harrowing	Rasmussen and Rasmussen (2000)
Denmark	Onion and cabbage	Manual weeding, torsion weeding, weed harrowing, Robovator	Melander et al. (2015)
Germany	Sugar beet	Manual steering, camera hoe, camera hoe with finger weeders, camera hoe with torsion weeder, camera hoe with rotary harrow	Kunz et al. (2016)
Germany	Soybean	Harrow, hoe	Weber et al. (2016)
Germany	Maize and sugar beet	Intelligent intrarow weed hoeing	Gerhards et al. (2016)
Germany	Sugar beet and soybean	Camera steering	Kunz et al. (2015)
Germany	Maize	Flexible tine harrow	Rueda-Ayala et al. (2015)
Germany	Sugar beet	Rotor tine	Rasmussen et al. (2012)
India	Soybean	Hand weeding	Singh et al. (2016)
Israel	Vegetables	Mechanical digger	Hershenhorn et al. (2015)
Italy	Pepper	Rotary hoeing	Campiglia et al. (2012)
Pakistan	Wheat	Bar harrow, hand hoeing	Jabran et al. (2012)
Pakistan	Rice	Hand weeding, hand hoeing	Akbar et al. (2011)
Pakistan	Rice	Interrow cultivation	Saqib et al. (2015)
Spain	Saffron	Flex tine harrow, finger weeder, torsion weeder	Cirujeda et al. (2014)
Spain	Cereals	Weed harrowing	Armengot et al. (2013)
Sweden	Willow	Torsion weeders	Albertsson et al. (2016)
The United Kingdom	Soybean	Steerage hoe	Tillett et al. (2002)
The United Kingdom	Cabbage	Steerage hoe equipped with conventional interrow cultivation blades	Tillett et al. (2008)
The United States	Bok choy, celery, lettuce, and radicchio	Rotating cultivar	Fennimore et al. (2014)
The United States	Cotton	Intrarow weeder	Saber et al. (2013)
The United States	Spring wheat	Harrowing, interrow cultivation	Kolb et al. (2012)
The United States	Corn	High-residue cultivator, vertical coulter/rotary harrow, high-residue rotary hoe	Bates et al. (2012)

the weed density by 29% compared with manual steering. It was concluded that automatic steering could be used to manage the sugar beet weeds with high efficacy.

8.7.2 Saffron

Cirujeda et al. (2014) evaluated different mechanical weeders for their weed control efficacy in saffron. Torsion and finger weeders, due to compacted soil, were difficult to adjust; however, more than 80% efficacy of weed control was observed during 2 out of 12 occasions where the weeders were tested. Flex tine harrow exceeded the 80% efficacy in 7 out of 18 occasions. Flex tine harrow did not require adjustments on the row and could be used in an aggressive position because of the 20 cm plantation depth of the corms. Weeds, including *Lolium rigidum* Gaud., could be effectively controlled with the spring tine harrow provided that climatic conditions allowed the tillage treatment immediately after flower harvest.

8.7.3 Soybean

Martelloni et al. (2016) studied the weed control efficacy of a spring tine harrow and an interrow cultivator in dry bean crop. They observed a significant decrease of 40% in weed density compared with control treatment with the interrow cultivator, whereas the decrease with the spring tine harrow was 14%. Weber et al. (2016) investigated the possibility of chemical and mechanical weed control strategies in soybean. The mechanical weed control gave comparable results as of herbicides. Herbicides resulted 92%–99% reduction in weed, whereas mechanical weed harrowing reduced weeds up to 82% and increased yield to 34% compared with control treatment.

8.8 CONCLUSIONS

Tillage is among the oldest methods of weed control. Nevertheless, the recent developments such as the manufacture of intelligent weeders prove that mechanical weeding is a pragmatic method of weed management. Certain limitations may limit the scope of mechanical weed control. Wise and timely implementation and the integration of mechanical weed control with other non-chemical weed management methods may help to achieve sustainable weed control. Most importantly, more research work is desired to limit the weaknesses of intelligent weeders and weed control robots. These may act as technologies of future environment-friendly weed control. In the future, robots may be assigned a duty of pulling weeds in crops or vegetables instead of humans (i.e., usually expensive).

References

Akbar, N., Jabran, K., Ali, M.A., 2011. Weed management improves yield and quality of direct seeded rice. Aust. J. Crop. Sci. 5, 688–694.

Albertsson, J., Verwijst, T., Rosenqvist, H., Hansson, D., Bertholdsson, N.O., Åhman, I., 2016. Effects of mechanical weed control or cover crop on the growth and economic viability of two short-rotation willow cultivars. Biomass Bioenergy 91, 296–305.

Alexandrou, A., 2004. Evaluation of In-Row Weed Cultivators in Organic Soybeans and Corn. Organic Farming Research Foundation, Wooster.

Ampong-Nyarko, K., De Datta, S.K., 1991. A Handbook for Weed Control in Rice. Int. Rice Res. Inst., Manila, Philippines.

Appleby, A.P., 2005. A history of weed control in the United States and Canada—a sequel. Weed Sci. 53, 762–768.

Armengot, L., José-María, L., Chamorro, L., Sans, F.X., 2013. Weed harrowing in organically grown cereal crops avoids yield losses without reducing weed diversity. Agron. Sustain. Dev. 33, 405–411.

Ascard, J., Bellinder, R.B.B., 1996. In: Mechanical in-row cultivation in row crops.Proceedings 1996 Second International Weed Control Congress, Copenhagen, Denmark, pp. 1121–1126.

Astrand, B., Baerveldt, A., 2002. An agricultural mobile robot with vision-based perception for mechanical weed control. Auton. Robot. 13, 21–35.

Baattrup-Pedersen, A., Larsen, S.E., Riis, T., 2002. Long-term effects of stream management on plant communities in two Danish lowland streams. Hydrobiologia 48, 33–45.

Baattrup-Pedersen, A., Larsen, S.E., Riis, T., 2003. Composition and richness of plant communities in small Danish streams—influence of environmental factors and weed cutting. Hydrobiologia 49, 171–179.

Baattrup-Pedersen, A., Riis, T., 2004. Impacts of different weed cutting practices on plant species diversity and composition in a Danish stream. River Res. Appl. 20, 103–114.

Bakker, T., 2009. An Autonomous Robot for Weed Control—Design, Navigation and Control. PhD Dissertation, Wageningen University, Wageningen.

Bates, R.T., Gallagher, R.S., Curran, W.S., Harper, J.K., 2012. Integrating mechanical and reduced chemical weed control in conservation tillage corn. Agron. J. 104, 507–517.

Baumann, D., 1992. In: Mechanical weed control with spring tine harrows (weed harrows) in row crops.Proceedings 1992 IXth International Symposium on the Biology of Weeds, Dijon, France, pp. 123–128.

Blackshaw, R.E., O'Donovan, J.T., Harker, K., Clayton, G.W., Stougaard, R.N., 2006. Reduced herbicide doses in field crops: a review. Weed Biol. Manage. 6, 10–17.

Bowman, G., 1997. Steel in the Field: A Farmer's Guide to Weed Management Tools. vol. 2 Sustainable Agriculture Network, Beltsville, MD.

Brandsæter, L.O., Mangerud, K., Rasmussen, J., 2012. Interactions between pre- and post weed harrowing in spring cereals. Weed Res. 52, 338–347.

Campiglia, E., Radicetti, E., Mancinelli, R., 2012. Weed control strategies and yield response in a pepper crop (Capsicum annuum L.) mulched with hairy vetch (Vicia villosa Roth.) and oat (Avena sativa L.) residues. Crop. Prot. 33, 65–73.

Chicouene, D., 2007. Mechanical destruction of weeds: a review. Agron. Sustain. Dev. 27, 19–27.

Chikowo, R., Faloya, V., Petit, S., Munier-Jolain, N.M., 2009. Integrated weed management systems allow reduced reliance on herbicides and long-term weed control. Agric. Ecosyst. Environ. 132, 237–242.

Christensen, S., Søgaard, H.T., Kudsk, P., Nørremark, M., Lund, I., Nadimi, E.S., Jørgensen, R., 2009. Site- specific weed control technologies. Weed Res. 49, 233–241.

Cirujeda, A., Melander, B., Rasmussen, K., Rasmussen, I.A., 2003. Relationship between speed, soil movement into the cereal row and intrarow weed control efficacy by weed harrowing. Weed Res. 43, 285–296.

Cirujeda, A., Aibar, J., Moreno, M.M., Zaragoza, C., 2015. Effective mechanical weed control in processing tomato: seven years of results. Renew. Agri. Food Syst. 30, 223–232.

Cirujeda, A., Marí, A.I., Aibar, J., Fenández-Cavada, S., Pardo, G., Zaragoza, C., 2014. Experiments on mechanical weed control in saffron crops in Spain. J. Plant Dis. Protect. 121, 223–228.

Cloutier, D.C., Weide, R.Y., van der Peruzzi, A., Leblanc, M.L., 2007. Mechanical Weed Management. CAB International, Wallingford, pp. 111–134.

Cordeau, S., Petit, S., Reboud, X., Chauvel, B., 2012. The impact of sown grass strips on the spatial distribution of weed species in adjacent boundaries and arable fields. Agric. Ecosyst. Environ. 155, 35–40.

Dawson, F.H., 1976. The annual production of the aquatic macrophyte Ranunculus penicillatus var. calcareus (R.W. Butcher) C.D.K. Cook. Aquat. Bot. 2, 51–73.

Dawson, F.H., 1989. Ecology and management of water plants in lowland streams. Rep. Freshwat. Biol. Ass. 57, 43–60.

De Wet, J.M., Harlan, J.R., 1975. Weeds and domesticates: evolution in the man-made habitat. Econ. Bot. 29, 99–108.

Farooq, M., Flower, K., Jabran, K., Wahid, A., Siddique, K.H., 2011. Crop yield and weed management in rainfed conservation agriculture. Soil Tillage Res. 117, 172–183.

Fennimore, S.A., Smith, R.F., Tourte, L., LeStrange, M., Rachuy, J.S., 2014. Evaluation and economics of a rotating cultivator in bok choy, celery, lettuce, and radicchio. Weed Technol. 28, 176–188.

Fogelberg, F., 2007. Reduction of manual weeding labour in vegetable crops—what can we expect from torsion weeding and harrowing?. In: Proceedings 2007 7th EWRS Workshop on Physical Weed Control, Salem, Germany, pp. 113–116. Available at: http://www.ewrs.org/pwc/proceedings.htm (accessed 28.03.08).

Fylkesmannen I Oslo Og Akershus, 2011. Tilskudd til regionale miljøtiltak for landbruket i Oslo og Akershus 2011. Landbruksavdelingen, Fylkesmannen i Oslo og Akershus, Oslo, Norway, 20 pp.

Gai, J., 2016. Plants Detection, Localization and Discrimination Using 3D Machine Vision for Robotic Intrarow Weed Control. Doctoral Dissertation, Iowa State University.

Gerhards, R., Sökefeld, M., Peteinatos, G., Nabout, A., Maier, J., Risser, P., 2016. Robotic intrarow weed hoeing in maize and sugar beet. Julius-Kühn-Archiv 452, 462–463.

Griepentrog, H.W., Nørremark, M., Nielsen, J., 2006. In: Autonomous intrarow rotor weeding based on GPS.Proceedings of the CIGR World Congress Agricultural Engineering for a Better World, Bonn, Germany. vol. 37. p. 17.

Hallefalt, F., Ascard, J., Olsson, R., 1998. Mechanical weed control by torsion weeder—a new method to reduce herbicide use in sugar beets. Asp. Appl. Biol. 52, 127–130.

Hamill, A.S., Holt, J.S., Mallory-Smith, C.A., 2004. Contributions of weed science to weed control and management 1. Weed Technol. 18, 1563–1565.

Hershenhorn, J., Zion, B., Smirnov, E., Weissblum, A., Shamir, N., Dor, E., Shilo, A., 2015. *Cyperus rotundus* control using a mechanical digger and solar radiation. Weed Res. 55, 42–50.

Inman, J.W., 2010. Into gear: success of the robocropweeder/thinner. Am. Veg. Grow. Available at: www.growingproduce.com.

Jabran, K., Ali, A., Sattar, A., Ali, Z., Yaseen, M., Iqbal, M.H.J., Munir, M.K., 2012. Cultural, mechanical and chemical weed control in wheat. Crop Environ. 3, 50–53.

Jeffries, M., Gannon, T., Yelverton, F., 2016. Herbicide inputs and mowing affects vaseygrass (*Paspalum urvillei*) control. Weed Technol. https://doi.org/10.1614/WT-D-16-00072.1.

Kareiva, P., Watts, S., McDonald, R., Boucher, T., 2007. Domesticated nature: shaping landscapes and ecosystems for human welfare. Science 316, 1866–1869.

Kern Hansen, U., Dawson, F.H., 1978. In: The standing crop of aquatic plants of lowland streams in Denmark and their inter-relationships of water nutrients in plant, sediment and water. EWRS 5th Symposium on Aquatic Weeds, pp. 143–150.

Kolb, L.N., Gallandt, E.R., Mallory, E.B., 2012. Impact of spring wheat planting density, row spacing, and mechanical weed control on yield, grain protein, and economic return in Maine. Weed Sci. 60, 244–253.

Korsmo, E., 1926. Ogras, Ograsarternas liv och kampen mot dem i nutidens jordbruk. Albert Bonniers Förlag, Stockholm, Sweden.

Kunz, C., Weber, J.F., Gerhards, R., 2016. Comparison of different mechanical weed control strategies in sugar beets. Julius-Kühn-Archiv 452, 446–451.

Kunz, C., Weber, J.F., Gerhards, R., 2015. Benefits of precision farming technologies for mechanical weed control in soybean and sugar beet—comparison of precision hoeing with conventional mechanical weed control. Agronomy 5, 130–142.

Kurstjens, D.A.G., Kropff, M.J., 2001. The impact of uprooting and soil-covering on the effectiveness of weed harrowing. Weed Res. 41, 211–228.

Kurstjens, D.A.G., 2002. Mechanisms of Selective Mechanical Weed Control by Harrowing. Dissertation, Wageningen University, Wageningen, the Netherlands, p. 200.

Loudon, J., 1871. The Horticulturist, Gardening in America Series. Applewood Books, p. 84. 9781429013680.

Lundkvist, A., 2009. Effects of pre- and post-emergence weed harrowing on annual weeds in peas and spring cereal. Weed Res. 49, 409–416.

Marshall, E.J.P., 1989. Distribution patterns of plants associated with arable field edges. J. Appl. Ecol. 26, 247–257.

Martelloni, L., Frasconi, C., Fontanelli, M., Raffaelli, M., Peruzzi, A., 2016. Mechanical weed control on small-size dry bean and its response to cross-flaming. Span. J. Agric. Res. 14, e0203.

Melander, B., 1997. Optimization of the adjustment of a vertical axis rotary brush weeder for intrarow weed control in row crops. J. Agric. Eng. Res. 68, 39–50.

Melander, B., Lattanzi, B., Pannacci, E., 2015. Intelligent versus non-intelligent mechanical intrarow weed control in transplanted onion and cabbage. Crop. Prot. 72, 1–8.

Merfield, C.N., 2014. The final frontier: non-chemical, intrarow, weed control for annual crops with a focus on mini-ridgers. Future Farm. Centre Bull. 4, 18.

Merfield, C.N., 2016. Robotic weeding's false dawn? Ten requirements for fully autonomous mechanical weed management. Weed Res. 56, 340–344.

Meyer, R.S., DuVal, A.E., Jensen, H.R., 2012. Patterns and processes in crop domestication: An historical review and quantitative analysis of 203 global food crops. New Phytol. 196, 29–48.

Milakovic, I., Karrer, G., 2016. The influence of mowing regime on the soil seed bank of the invasive plant *Ambrosia artemisiifolia* L. NeoBiota 28, 39–49.

Nørremark, M., Griepentrog, H.-W., 2004. In: Analysis and definition of the close-to-crop area in relation to robotic weeding.Presented at the 6th Workshop of the EWRS Working Group "Physical and Cultural Weed Control," Lillehammer, Norway. Retrieved from: http://orgprints.org/4834.

Pannacci, E., Lattanzi, B., Tei, F., 2017. Non-chemical weed management strategies in minor crops: a review. Crop. Prot. 96, 44–58.

Pannacci, E., Tei, F., 2014. Effects of mechanical and chemical methods on weed control, weed seed rain and crop yield in maize, sunflower and soyabean. Crop. Prot. 64, 51–59.

Peruzzi, A., Ginanni, M., Raffaelli, M., Fontanelli, M., 2005. In: Physical weed control in organic carrots in the Fucino Valley, Italy.Proceedings of 13th EWRS Symposium, Bari, June, pp. 19–23.

Pitlo, R.H., Dawson, F.H., 1990. Flow-resistance of aquatic weeds. In: Pieterse, A.H., Murphy, K.J. (Eds.), Aquatic Weeds. The Ecology and Management of Nuisance Aquatic Vegetation. Oxford University Press, Oxford.

Prince, J.M., Shaw, D.R., Givens, W.A., Newman, M.E., Owen, M.D.K., Weller, S.C., Young, B.G., Wilson, R.G., Jordan, D.L., 2012. Benchmark study: III. Survey on changing herbicide use patterns in glyphosate-resistant cropping systems. Weed Technol. 26, 536–542.

Rasmussen, J., Ascard, J., 1995. Weed control in organic farming systems. In: Glen, D.M., Greaves, M.P., Anderson, H.M. (Eds.), Ecology and Integrated Farming Systems. Wiley, Chichester, UK, pp. 46–67.

Rasmussen, K. (2002). Weed control by a rolling cultivator in potatoes. In: Proceedings 2002 5th EWRS Workshop on Physical and Cultural Weed Control, Pisa, Italy, pp. 111–118. Available at: http://www.ewrs.org/pwc/archive.htm (accessed 28.03.08).

Rasmussen, I.A., 2004. The effect of sowing date, stale seedbed, row width and mechanical weed control on weeds and yields of organic winter wheat. Weed Res. 44, 12–20.

Rasmussen, J., Griepentrog, H.W., Nielsen, J., Henriksen, C.B., 2012. Automated intelligent rotor tine cultivation and punch planting to improve the selectivity of mechanical intra-row weed control. Weed Res. 52, 327–337.

Rasmussen, K., Rasmussen, J., 2000. Barley seed vigour and mechanical weed control. Weed Res. Oxford 40, 219–230.

Rueda-Ayala, V., Peteinatos, G., Gerhards, R., Andújar, D., 2015. A non-chemical system for online weed control. Sensors 15, 7691–7707.

Sabanci, K., Aydin, C., 2017. Smart robotic weed control system for sugar beet. J. Agric. Sci. Technol. 19, 73–83.

Saber, M.N., Lee, W.S., Burks, T.F., MacDonald, G.E., Salvador, G.A., 2013. An automated mechanical weed control system for organic row crop production. American Society of Agricultural and Biological Engineers, Kansas City, MO, p. 1.

Saqib, M., Ehsanullah, N.A., Latif, M., Ijaz, M., Ehsan, F., Ghaffar, A., 2015. Development and appraisal of mechanical weed management strategies in direct seeded aerobic rice (*Oryza sativa* L.). Pak. J. Agric. Sci. 52, 587–593.

Singh, M., Bhullar, M.S., Chauhan, B.S., 2015. Seed bank dynamics and emergence pattern of weeds as affected by tillage systems in dry direct-seeded rice. Crop. Prot. 67, 168–177.

Singh, M., Dudwe, T.S., Verma, A.K., 2016. Integration of chemical and mechanical weed management to enhance the productivity of soybean (*Glycine max*). J. Progressive Agri. 7, 132–135.

Slaughter, D.C., Giles, D.K., Downey, D., 2008. Autonomous robotic weed control systems: a review. Comput. Electron. Agric. 61, 63–78.

Tarasoff, C.S., Streichert, K., Gardner, W., Heise, B., Church, J., Pypker, T.G., 2016. Assessing benthic barriers vs. aggressive cutting as effective yellow flag iris (*Iris pseudacorus*) control mechanisms. Invas. Plant Sci. Manage. 9, 229–234.

Tillett, N.D., Hague, T., Miles, S.J., 2002. Interrow vision guidance for mechanical weed control in sugar beet. Comput. Electron. Agric. 33, 163–177.

Tillett, N.D., Hague, T., Grundy, A.C., Dedousis, A.P., 2008. Mechanical within-row weed control for transplanted crops using computer vision. Biosyst. Eng. 99, 171–178.

Timmons, F., 1970. A history of weed control in the United States and Canada. Weed Sci. 18, 294–307.

Timmons, F., 2005. A history of weed control in the United States and Canada. Weed Sci. 53, 748–761.

Tu, M., Hurd, C., Randall, J.M., 2001. Weed Control Methods Handbook: Tool and Techniques for Use in Natural Areas. The Nature Conservancy, 2001. Available from: http://digitalcommons.usu.edu/cgi/viewcontent.cgi?article=1532&context=govdocs (accessed July 6, 2017).

USAID, 1984. Annual Progress Report. United States Agency for International Development.

van der Schans, D., Bleeker, P., Molendijk, L., Plentinger, M., Van der Weide, R., Lotz, B., Bauermeister, R., Total, R., Baumann, D.T., 2006. Practical Weed Control in Arable Farming and Outdoor Vegetable Cultivation without Chemicals. PPO Publication 532, Applied Plant Research, Wageningen University, Lelystad, The Netherlands, p. 77.

Van der Weide, R.Y., Kurstjens, D., 1996. Eginstelling en selectiviteit. Ekoland 16, 14–15.

Van der Weide, R.Y., Bleeker, P.O., Achten, V.T.J.M., Lotz, L.A.P., Fogelberg, F., Melander, B., 2008. Innovation in mechanical weed control in crop rows. Weed Res. 48, 215–224.

Weber, J.F., Kunz, C., Gerhards, R., 2016. Chemical and mechanical weed control in soybean (*Glycine max*). Julius-Kühn-Archiv 452, 171–176.

Westerdijk, C.E., Van Der Weide, R.Y., Wevers, J.A., 1997. In: Integrated weed control (harrowing) in sugar beet.Proceedings of the 60th IIRB Congress," Cambridge, UK, pp. 579–584.

Wiltshire, J., Tillett, N., Hague, T., 2003. Agronomic evaluation of precise mechanical hoeing and chemical weed control in sugar beet. Weed Res. 43, 236–244.

Young, S.L., Meyer, G.E., Woldt, W.E., 2014. Future directions for automated weed management in precision agriculture. In: Young, S.L., Pierce, F.J. (Eds.), Automation: The Future of Weed Control in Cropping Systems. Springer Netherlands, Dordrecht, pp. 249–259. Retrieved from: http://link.springer.com/10.1007/978-94-007-7512-1_15.

Zohary, D., Hopf, M., Weiss, E., 2012. Domestication of Plants in the Old World: The Origin and Spread of Domesticated Plants in Southwest Asia, Europe, and the Mediterranean Basin. Oxford University Press on Demand, Oxford.

Further Reading

Fogelberg, F., Gustavsson, A.D., 1999. Mechanical damage to annual weeds and carrots by in-row brush weeding. Weed Res. 39, 469–479.

Kouwenhoven, J.K., 1997. Intrarow mechanical weed control—possibilities and problems. Soil Tillage Res. 41, 87–104.

Index

Note: Page numbers followed by *f* indicate figures, and *t* indicate tables.